When the Continuum Breaks Down: Unveiling the Limits of Classical Fluid Mechanics

Robirt

Contents

Chapter 1

Introduction

1.1 Motivation

In classical fluid mechanics, fluid structure at the microscopic level is largely ignored, due to the fact that the molecular nature of fluids exists on scales which are much smaller than the dimensions that are traditionally of interest. The length scales associated with the molecular effects of a gas can be characterized by defining the mean free path of fluid molecules, λ. This is the average distance that gas molecules travel between inter-particle collisions. It is often used to classify the regime of a gas flow through definition of the Knudsen number,

$$\text{Kn} = \frac{\lambda}{\ell}. \tag{1.1}$$

This is the ratio of the mean free path to a characteristic length for the situation, ℓ. Gas flow regimes are then divided according to this non-dimensional number:

$$
\begin{aligned}
\text{Kn} &< 0.01 &\quad &\text{continuum regime,} \\
0.01 \leq \text{Kn} &< 10 &\quad &\text{transition regime,} \\
10 \leq \text{Kn} & &\quad &\text{free-molecular regime.}
\end{aligned}
$$

Classical fluid mechanics relies on the continuum assumption, which ultimately depends on very low Knudsen numbers of $\text{Kn} < 0.01$. In this regime, the scales associated with molecular processes are much smaller than the representative length for the situation, and traditional fluid-mechanic techniques can provide valid descriptions of the flow. Changes in fluid properties here are well described by the compressible Euler or Navier-Stokes equations. However, when the Knudsen number becomes larger, these techniques are no longer valid, and a treatment of gas-particle behaviour becomes important for describing the microscopic nature of the fluid. For the transition regime, taking moments of the Boltzmann equation can lead to valid descriptions of non-equilibrium, micro-scale flows.

These derived models offer many numerical advantages, due to their hyperbolic, first-order structure. However, these descriptions also contain stiff source terms, characterized by a relaxation time. Due to the time-scale of this value, time-step values in explicit numerical methods are severely limited in order to achieve both stability and accuracy. An implicit treatment helps to overcome this restriction, but a larger time step obtained in this fashion does not guarantee accuracy. The current study focuses on a known method that was designed specifically to accurately and efficiently solve hyperbolic models with stiff relaxation source terms derived from the kinetic theory of gases.

1.2 Objectives of the Current Study

Moment methods derived from the kinetic theory of gases are hyperbolic, first-order PDEs with stiff local source terms of the form

$$\frac{\partial U}{\partial t} + \frac{\partial F_i}{\partial x_i} = S. \tag{1.2}$$

Here, U is the solution vector, F_i is the flux dyad, and S is the local source term. The main objective of this study is to implement a numerical method that can accurately solve models of this form. A typical finite-volume method (FVM) is often used for this purpose. In such methods, cell-centered data is used to represent the solution over the computational domain, and achieving higher-order spatial accuracy relies on piecewise-polynomial reconstruction within cells. However, this requires extending the stencils used for data representation in each cell. Popular higher-order FVM schemes include van Leer's MUSCL (Monotone Upstream-centered scheme for Conservation Laws) [24], and Colella & Woodward's PPM (Piecewise Parabolic Method) [10], which are second-order and third-order accurate in space respectively. In order to ensure stability and solution monotonicity for such schemes, a total-variation-diminishing (TVD) limiter is required. A separate class of methods, known as discontinuous-Galerkin methods, avoid the issue of growing stencils entirely. Instead of a piecewise-constant solution in each cell, DG schemes use a polynomial of degree k to represent the solution. A visual comparison of the solution representations used by a DG scheme and that of a first- and second-order finite-volume method is shown in Figure 1.1. A specific DG scheme is of particular interest for this study. The discontinuous Galerkin Hancock (DGH) method originally proposed by Suzuki and van Leer [38] is based upon Huynh's "upwind moment scheme" [17]. The DGH method extends Huynh's model for hyperbolic conservation laws to hyperbolic-relaxation equations, and is able to achieve third-order accuracy while using only linear elements in each cell. Two key characteristics for this model are:

1. cell variables at flux quadrature points are updated over a half-time step without any interactions with neighbouring cells (Hancock's technique [1]),

2. the gradient of each flow variable evolves as described by an independent equation (DG representation).

The application of this scheme to moment methods and other hyperbolic PDEs yields a highly efficient and accurate numerical treatment for viscous, compressible gas flows both in and out of local thermodynamic equilibrium.

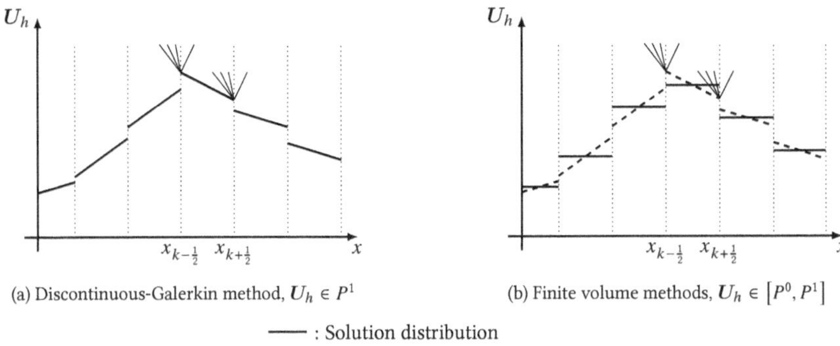

(a) Discontinuous-Galerkin method, $U_h \in P^1$ (b) Finite volume methods, $U_h \in \left[P^0, P^1\right]$

Figure 1.1: Solution distribution over a domain for a DG-P^1 method (a), and a first- and second-order finite volume method (b).

Up until now, the DGH method has never been used for the solution of hyperbolic-relaxation PDEs on a large-scale. The objective of the current work was to implement the scheme in a parallel framework for use with modern distributed clusters. The major, original contributions shown in this work include:

- First large-scale, general implementation of the DGH method for use in one, two, and three dimensions.

- Implementation of block-based adaptive mesh refinement used for dynamically improving the solution of a range of hyperbolic-relaxation conservation laws during time-marching.

- The first convergence study for the DGH method on a non-Cartesian, multi-dimensional mesh for a non-linear set of PDEs.

- Strong scaling analyses for parallel computations performed using the current implementation of the method.

1.3 Scope of the Current Study

This book covers the formulation and implementation of the discontinuous-Galerkin Hancock method. In Chapter 2, a review of the kinetic theory of gases is presented, as well as the concept of moment closures. Next, in Chapter 3, the DGH method is introduced. The weak formulation of a DG method is discussed, which leads into the three-dimensional construction of the scheme. This section concludes with a stability analysis of the scheme. Chapter 4 discusses the parallel implementation for adaptive mesh refinement and load balancing across distributed machines. Numerical results obtained using the DGH scheme for many different PDEs are presented in Section 5.1, as well as parallel efficiency studies.

Chapter 2

Moment Methods

2.1 Kinetic Theory of Gases

Gas flows can be thought of as a collection of particles interacting with each other. Using Newton's law of motion and classical mechanics, collisions and other transfers of energy within the flow can be described mathematically. However, constructing a set of conservation equations in this fashion for every single particle in a flow is prohibitively expensive. To overcome this issue, the kinetic theory of gases relies on a statistical description of the group of fluid particles, rather than each particle individually. The classical kinetic theory of gases is based upon the following assumptions:

- All matter is composed of discrete particles.

- All particles of a given species are identical.

- Particles are simply points, with no internal structure.

- Particles only exert forces over distances much smaller than the mean-free path.

- Only binary collisions occur between particles, meaning that at most two particles interact in a given collision.

- Quantum effects are neglected.

- Colliding particles are statistically uncorrelated.

More modern extensions of the kinetic theory that relax these assumptions do exist. However, they are not germane to the present work.

2.2 Velocity Distribution Functions

In gas-kinetic theory, it is assumed that a gas is comprised of discrete particles. However, rather than model the behaviour of these particles individually, the ensemble of particles is treated statistically. A probability distribution function, $\mathcal{F}(x_i, v_i, t)$, is defined in phase space, which is the combination of physical space and velocity space. Looking at a cube in phase space with volume $dx_i dv_i$, such that

$$dx_i dv_i = dxdydzdv_x dv_y dv_z, \tag{2.1}$$

the number of particles within this cube at a location, x_i, with velocity, v_i, can be found as

$$N_{x_i,v_i}(t) = \mathcal{F}(x_i, v_i, t)dx_i dv_i. \tag{2.2}$$

The number density of particles in a system is therefore

$$n(x_i, t) = \iiint_{-\infty}^{\infty} \mathcal{F}(x_i, v_i)\, dv_i. \tag{2.3}$$

2.3 Moments of Distribution Functions

Moments provide a connection between the distribution function and macroscopic properties. To take a moment of a distribution function, the function is multiplied by an appropriate velocity-dependent weight, $W(v_i)$, and integrated over all velocity space. For example, the mass density, ρ, can be found by taking $W(v_i) = m$, where m is the mass of a particle. This is expressed as

$$\rho = \iiint_{-\infty}^{\infty} m\mathcal{F}(x_i, v_i)\, dv_i = \langle m\mathcal{F}\rangle. \tag{2.4}$$

The shorthand notation, $\langle W(v_i)\mathcal{F}\rangle$, is used to denote integration over all velocity space. Similarly, the momentum density of the gas is found by taking $W(v_i) = mv_i$, such that

$$\rho u_i = \langle mv_i\mathcal{F}\rangle. \tag{2.5}$$

Using the definitions of mass and momentum density, the bulk velocity of the fluid can be found with

$$u_i = \frac{\langle mv_i\mathcal{F}\rangle}{\langle m\mathcal{F}\rangle}. \tag{2.6}$$

Once the bulk velocity is found, the particle velocity, v_i, can be written as the sum of two components: the bulk velocity, u_i, and the random velocity, c_i, such that $v_i = u_i + c_i$. Moments can also be taken using weights dependent on the random velocity, i.e. $W(c_i)$. For example, the second-order pressure tensor, P_{ij}, is defined as

$$P_{ij} = \langle mc_i c_j\mathcal{F}\rangle. \tag{2.7}$$

This tensor is symmetric, with six distinct entries. For a monatomic gas, it is related to the traditional thermodynamic pressure with $p = P_{ii}/3$, and to the deviatoric (viscous) stress tensor with $\tau_{ij} = p\delta_{ij} - P_{ij}$. Here, δ_{ij} is the Kronecker delta, defined as

$$\delta_{ij} = \begin{cases} 1, & \text{if } i = j, \\ 0, & \text{otherwise.} \end{cases} \tag{2.8}$$

Moments of arbitrarily high order can be taken, however the physical significance of a moment becomes more ambiguous as this order is increased.

2.4 The Maxwell-Boltzmann Distribution

Local thermodynamic equilibrium is the state to which any gas, when left alone, will relax due to inter-particle collisions. The distribution function describing this state is defined as the Maxwell-Boltzmann distribution,

$$\mathcal{M} = n \left(\frac{\beta(x_i, t)}{\pi} \right)^{\frac{3}{2}} e^{-\beta(x_i, t) v_i v_i}, \tag{2.9}$$

where

$$\beta(x_i, t) = \frac{m}{2kT(x_i, t)}. \tag{2.10}$$

Here, $k = 1.3806 \times 10^{-23}$ J/K is the Boltzmann constant, and $T(x_i, t)$ is the temperature of the fluid at location x_i and time t. It has been proven that a gas in any arbitrary non-equilibrium state will be driven towards the Maxwell-Boltzmann distribution due to particle collisions. Once it reaches this state, particle collisions will no longer have any effect on the distribution function.

2.5 The Boltzmann Equation

The evolution of the distribution function for a gas is described by the Boltzmann equation,

$$\frac{\partial \mathcal{F}}{\partial t} + v_i \frac{\partial \mathcal{F}}{\partial x_i} + \frac{\partial}{\partial v_i} (a_i \mathcal{F}) = \frac{\delta \mathcal{F}}{\delta t}. \tag{2.11}$$

Here, a_i is the particle acceleration due to external forces, which is taken to be zero for now. The right hand side term, $\frac{\delta \mathcal{F}}{\delta t}$, is known as the collision operator. This term models the effects of inter-particle collisions on the distribution function.

2.5.1 Boltzmann Collision Operator

In order to treat the collision operator, a few assumptions are made:

1. only binary collisions take place, on a scale much smaller than the mean free path,

2. over this range, the distribution function can be assumed to be constant,

3. it takes many collisions to appreciably change $\mathcal{F}$; it is approximately constant through one collision,

4. particle velocities are statistically independent.

Instead of resolving the time-scale of a collision, it is imagined that pre-collision particles with velocities (v_i, v_i^1) are instantly removed, and post-collision particles with velocities $(v_i', v_i^{1'})$ are added. Following these assumptions, the collision operator can be written as

$$\frac{\delta \mathcal{F}}{\delta t} = \left\langle \int_0^{2\pi} \int_0^{\pi} \left(\mathcal{F}' \mathcal{F}^{1'} - \mathcal{F} \mathcal{F}^1 \right) g\sigma \sin\chi \, \mathrm{d}\chi \mathrm{d}\epsilon \right\rangle_{v_i'} . \tag{2.12}$$

In this expression, $\mathcal{F}, \mathcal{F}^1, \mathcal{F}', \mathcal{F}^{1'}$ are the distribution function evaluated at $v_i, v_i^1, v_i', v_i^{1'}$. Here, the relative speed of the particles prior to the collision is denoted as $g = |v_i - v_i^1|$, σ is the differential collision cross section, χ is the deflection angle, and ϵ is an angle describing the orientation of the collision plane. The collision operator is high-dimensional, and complicated to evaluate in general cases. Simplified models for collisions are therefore often used; one of the most common approximations is the BGK operator [6], written as

$$\frac{\delta \mathcal{F}}{\delta t} = -\frac{\mathcal{F}(x_i, v_i, t)}{\tau_{\mathcal{F}}(x_i, t)} + \frac{\mathcal{M}(x_i, v_i, t)}{\tau_{\mathcal{M}}(x_i, t)}. \tag{2.13}$$

For this model, particles in non-equilibrium states are removed, while particles in equilibrium are added using characteristic time scales $\tau_{\mathcal{F}}$ and $\tau_{\mathcal{M}}$ respectively. To ensure conservation of particles, these times are often taken to be equal, such that $\tau_{\mathcal{F}} = \tau_{\mathcal{M}} = \tau$, reducing Equation (2.13) to

$$\frac{\delta \mathcal{F}}{\delta t} = -\frac{\mathcal{F}(x_i, v_i, t) - \mathcal{M}(x_i, v_i, t)}{\tau}. \tag{2.14}$$

Values for the characteristic relaxation time τ can be selected to produce models in agreement with macroscopic transport coefficients. For example, the equilibrium fluid viscosity can be recovered by choosing $\tau = \mu/p$. For typical gases at standard pressure, this leads to relaxation times on the order of 10×10^{-10} s. As can be seen, when thermodynamic equilibrium is reached and $\mathcal{F} = \mathcal{M}$, this operator will be equal to zero, as needed.

2.5.2 Maxwell's Equation of Change

In order to derive equations for the evolution of chosen macroscopic properties, a set of weights are used for directly taking moments of the Boltzmann equation. For an arbitrary velocity-dependent weight W, this yields

$$\frac{\partial}{\partial t} \langle mW\mathcal{F} \rangle + \frac{\partial}{\partial x_i} \langle mv_i W\mathcal{F} \rangle = \Delta \left(W\mathcal{F} \right), \tag{2.15}$$

where $\Delta \left(W\mathcal{F} \right) = \langle mW\frac{\delta \mathcal{F}}{\delta t} \rangle$. For a desired set of macroscopic properties, there exists a set of related weights,

$$W = [W_0, W_1, \ldots, W_N]^\top . \tag{2.16}$$

The resulting vector of moments can now be defined as

$$U = \langle m W \mathcal{F} \rangle,$$ (2.17)

where U is the solution vector containing conserved variables. Rewriting Equation (2.15) with these definitions yields

$$\frac{\partial U}{\partial t} + \frac{\partial}{\partial x_i} \langle m v_i W \mathcal{F} \rangle = \Delta (W \mathcal{F}).$$ (2.18)

It is appealing to rewrite Equation (2.18) in balance law form by defining the flux dyad $F_i = \langle m v_i W \mathcal{F} \rangle$ and local source term $S = \Delta (W \mathcal{F})$, yielding

$$\frac{\partial U}{\partial t} + \frac{\partial F_i}{\partial x_i} = S$$ (2.19)

2.6 Moment Closure

Solving for or approximating a solution to the seven-dimensional Boltzmann equation can be done a number of ways. Discretizing the Boltzmann equation directly requires doing so for both the spatial domain of the problem, as well as a sufficiently large domain of velocity space. The cost of this approach is prohibitively expensive, as is direct treatment of each gas particle. Another commonly used method is the Direct Simulation Monte Carlo (DSMC) method [7], which randomly chooses a subset of particles and time marches them using probabilistic models. This method is valid, but converges very slowly due to statistical noise.

The method of moment closures can provide alternative models for describing the evolution of gases. For many purposes, fluid properties at the microscopic level are not important; it is usually sufficient to obtain solutions for a subset of macroscopic properties. The original moment closures for non-equilibrium gases were proposed by Grad, where he derived thirteen- and twenty-moment closures [15].

Equation (2.19) represents a set of first-order PDEs describing the evolution of macroscopic properties of a fluid. However, there is still an issue here. The flux dyad and source term cannot, in general, be determined as functions of the variables present in the solution vector, and as a result, the system is not closed. Due to the structure of this equation, the flux dyad will always have moments of one order higher than those in the solution vector. In order to close the system, $\mathcal{F}$ is usually restricted to a prescribed form. This assumed form should contain the same number of free parameters, $\alpha = [\alpha_1, \alpha_2, \ldots, \alpha_N]^\top$ as the number of entries in U. The values for these parameters are chosen such that the previous moment relationships shown in Equations (2.4)–(2.7) are satisfied . Since the assumed form of $\mathcal{F}$ is then known, any higher-order moment that appears in the flux dyad can then be directly integrated and the system will be closed.

In Grad's original closures, a non-equilibrium distribution function was approximated as a polynomial expansion around local equilibrium, such that

$$\mathcal{F}_{\text{Grad}} = \mathcal{M} (\alpha^\top W).$$ (2.20)

Because this assumed distribution is simply a product between a polynomial and the Maxwellian distribution, it is not strictly positive. This means the model could predict a negative number of particles at a given location with a certain velocity. The other, more devastating problem, is that the flux Jacobian for models of this form can develop complex eigenvalues, leading to a loss of hyperbolicity for the resulting PDEs. Mathematically, this model cannot produce a solution for many realistic problems.

2.7 Maximum-Entropy Moment Closures

Following Grad's original moment closures, alternate techniques have been used for assuming the form of the distribution function. A newer hierarchy of closures assumes a form for $\mathcal{F}$ which maximizes entropy, while at the same time satisfying a given set of velocity moments present in the solution vector U [26, 29]. Boltzmann proved that the entropy density for a non-equilibrium gas can be found as

$$S = \left\langle -k\mathcal{F} \ln \frac{\mathcal{F}}{y} \right\rangle, \tag{2.21}$$

where y is a constant used to non-dimensionalize $\mathcal{F}$. To derive the form of distribution function that maximizes entropy, an associated velocity weight is defined such that

$$M = -\frac{k}{m} \left(\ln \frac{\mathcal{F}}{z} - 1 \right). \tag{2.22}$$

Here, $y = ez$. The problem of maximizing entropy while satisfying all moments in U, corresponding to velocity weights W is therefore

$$\max_{\mathcal{F}} \left\langle -k \left(\mathcal{F} \ln \frac{\mathcal{F}}{z} - \mathcal{F} \right) \right\rangle. \tag{2.23}$$

Using Lagrange multipliers, this constrained maximization problem can be solved. A vector of Lagrange multipliers, $\hat{\alpha}$, is defined, and the maximization problem becomes a search for the critical points of a function, J, of the form

$$J = \left\langle -k \left(\mathcal{F} \ln \frac{\mathcal{F}}{z} - \mathcal{F} \right) \right\rangle - \hat{\alpha} \left(U - \langle W\mathcal{F} \rangle \right). \tag{2.24}$$

At a critical point, $\frac{\mathrm{d}J}{\mathrm{d}\mathcal{F}} = 0$, yielding

$$\frac{\mathrm{d}J}{\mathrm{d}\mathcal{F}} = \left\langle -k \ln \frac{\mathcal{F}}{z} \right\rangle - \hat{\alpha} \langle W \rangle \tag{2.25}$$

$$0 = \langle \ln \mathcal{F} - \alpha W \rangle.$$

Here, k and z have been absorbed into the Lagrange multipliers. Clearly, Equation (2.25) is satisfied if the distribution function is of the form

$$\mathcal{F} = e^{\alpha^\top W}. \tag{2.26}$$

It can be seen that the closure coefficients are the Lagrange multipliers obtained from the constrained maximization problem.

Because the polynomial $\alpha^\top W$ is now in an exponential, this assumed form is guaranteed to be positive. There are also strong physical arguments as to why models derived in this fashion should provide accurate predictions. Particle collisions will always cause entropy to increase, and the maximum-entropy distribution is statistically the most likely distribution which is consistent with the known moments. The last bit of good news is that maximum entropy moment closures lead to globally hyperbolic moment equations whenever a distribution function of the form given in Equation (2.26) that agrees with the known moments exists. However, Junk proved that for some high-order moment methods, there are valid, physical gas states for which corresponding distribution functions of this form do not exist [19, 20].

Two low-order members of this hierarchy that are commonly used are the five-moment (Euler) and ten-moment (Gaussian) closures. These models are derived using weight vectors ($W_5 = [m, mv_i, mv_iv_i]^\top$, $W_{10} = [m, mv_i, mv_iv_j]^\top$) respectively. The Gaussian closure is well suited to viscous, adiabatic flows, despite not having a treatment for heat flux. The lowest-order member of the maximum-entropy hierarchy which does have a treatment for heat flux is a fourteen-moment closure [28]. It is generated with weights $W_{14} = [m, mv_i, mv_iv_j, mv_iv^2, mv^4]^\top$. The low-order moment closures have proven to be well suited for predicting both equilibrium and non-equilibrium flows. The higher-order moment closures are promising, but present difficult numerical problems due to the fact that their distribution function cannot be integrated directly. As a result, there is no closed form for the fluxes of the system, and Lagrange multipliers must be found using a computationally expensive, ill-conditioned iterative process. To now, this has restricted the adoption of higher-order moment methods.

Chapter 3

The Discontinuous-Galerkin Hancock Method

3.1 Introduction

Moment methods derived from the kinetic theory of gases provide models with physical and mathematical advantages over classical techniques. A spatial discretization such as a standard finite-volume method is often used for the numerical solution of such PDEs. The discontinuous-Galerkin method is a strong alternative that allows for high-order spatial accuracy on arbitrary structured or unstructured computational domains, and the potential for high parallel efficiency.

The finite-element discontinuous-Galerkin method was first presented by Reed and Hill as a technique used for solving the steady linear neutron transport equation [33]. Following this, LeSaint and Raviart shared error analyses showing that a DG method of degree k has a rate of convergence of $(\Delta x)^k$ for general triangulations and of $(\Delta x)^{k+1}$ for Cartesian grids [25]. Johnson and Pitkäranta later proved the rate of convergence for general triangulations is $(\Delta x)^{k+\frac{1}{2}}$ [18], which was afterwards confirmed by Peterson [31]. An example of a popular model in this family is the Runge-Kutta discontinuous-Galerkin (RKDG) method, primarily developed by Cockburn and Shu [9], which originally used a piecewise linear DG method for its spatial discretization, and an explicit TVD second-order Runge-Kutta scheme for temporal discretization. In recent years, higher-order polynomials and Runge-Kutta methods have been used to extend the model to higher orders of accuracy.

Because RKDG schemes adopt DG discretization in space only, a semi-discrete method is applied for the temporal discretization. A semi-discrete method using the method-of-lines (MOL) to separate the spatial and temporal discretizations was first described by Schiesser [36]. This method approximates a PDE with a system of ODEs, which can be advantageous, but it is not without its drawbacks. The reduced stability of this approach results in severe time step limitations, which become even more restrictive as the DG method used for spatial discretization increases in order of accuracy.

A fully discrete method for space-time discretization can be employed to overcome the shortcomings associated with the MOL approach. A classical example of a fully discrete method is the

Lax-Wendroff method, which is second-order in both space and time [22]. This method expresses temporal derivatives in terms of the spatial derivatives. A more recent example is the arbitrarily high order (ADER) method, developed by Toro *et al.* [39], and later applied to a DG framework by Dumbser and Munz [11]. The ADER-DG method is one-step, and more efficient than a multi-stage RKDG method due to reduced use of a TVD limiter. However, a Fourier analysis proved that the fully discrete ADER-DG method has a similar stability restriction as RKDG methods [12].

The discontinuous-Galerkin Hancock model, developed by Suzuki and Van Leer [38], utilizes a coupled space-time approach, but can be seen as being in between semi-discrete and fully discrete due to the fact that the test functions for solution representation are only dependent on space. This method is based upon Huynh's "upwind moment scheme" [17], originally developed for the numerical solution of hyperbolic conservation laws, and extends the scheme to hyperbolic-relaxation equations. The upwind moment scheme is a one-step, fully discrete method, with an intermediate step needed to compute the volume integral of the fluxes. The scheme requires two Riemann problems be solved at each cell interface in the computational domain, and is third-order accurate in both space and time. Due to the large amount of local work required and low amount of communication with neighbours, the discontinuous-Galerkin Hancock method is well suited to parallel computations.

3.2 Formulation of the DGH Method for Three-Dimensional Equations

3.2.1 Weak Formulation

The DGH method is a coupled space-time method for the solution of hyperbolic balance laws with stiff local source terms. Specifically, weak solutions for PDEs of the form

$$\frac{\partial U}{\partial t} + \frac{\partial F_i}{\partial x_i} = S, \tag{3.1}$$

are sought such that the product of the PDE with a chosen, finite set of test functions, integrated over the space-time domain $\Omega(t) \times T$ should be satisfied. A scalar test function is defined in the same space as the solution $U(x_j, t)$, denoted as $v(x_j, t) \in \Omega(t) \times T$. By taking the product of the PDE with an arbitrary test function and integrating over the space-time domain, the weak solution of the PDE is obtained. This can be written as

$$\iint\limits_{\Omega(t) \times T} v(x_j, t) \frac{\partial U}{\partial t} \, dx_j dt = - \iint\limits_{\Omega(t) \times T} v(x_j, t) \frac{\partial F_i}{\partial x_i} \, dx_j dt + \iint\limits_{\Omega(t) \times T} v(x_j, t) S \, dx_j dt. \tag{3.2}$$

Here, T is the time interval over $\left[t^n, t^{n+1}\right]$. The weak formulation of the problem is then:

Find $U(x_j, t)$ such that Equation (3.2) is satisfied for a chosen set of test functions $v(x_j, t) \in \Omega(t) \times T$.

Test functions are non-zero in only one cell, and the spatial domain of integration $\Omega(t)$ can be reduced to that of one cell, $\Omega(t)_k$. Integration by parts is then applied to Equation (3.2), which removes derivatives of the solution and its flux from the space-time integrals,

$$\int_T v(x_j,t)\frac{\partial U}{\partial t}\,dt = \left[U^{n+1}v(x_j,t) - U^n v(x_j,t)\right] - \int_T U\frac{\partial}{\partial t}v(x_j,t)\,dt, \tag{3.3a}$$

$$\int_{\Omega(t)_k} v(x_j,t)\frac{\partial F_i}{\partial x_i}\,dx_j = \int_{\Gamma_k} v(x_j,t)F_i \cdot \hat{n}_i\,d\Gamma - \int_{\Omega(t)_k} F_i\frac{\partial}{\partial x_i}v(x_j,t)\,dx_j. \tag{3.3b}$$

Here, Γ_k is the boundary of the three-dimensional cell, and $\hat{n}_i$ is the outward unit normal to the boundary. When it is assumed that the spatial domain over the time interval T is fixed (i.e. $\Omega(t) = \Omega$) and the test function is a function of space only, $v(x_j,t) = v(x_j)$, the weak formulation can be further simplified. By substituting Equations (3.3a) and (3.3b) into Equation (3.2) and applying Fubini's theorem, which allows for the alternation of the order for integration in space and time, the weak formulation becomes

$$\int_{\Omega_k} v(x_j)\left[U^{n+1} - U^n\right]\,dx_j = -\iint_{\Gamma_k \times T} v(x_j)F_i \cdot \hat{n}_i\,d\Gamma dt + \iint_{\Omega_k \times T} F_i\frac{\partial}{\partial x_i}v(x_j)\,dx_j dt$$
$$+ \iint_{\Omega_k \times T} v(x_j)S\,dx_j dt. \tag{3.4}$$

3.2.2 Polynomial Representation of the Solution

The individual solution $U_h(x_j,t)|_{\Omega_k}$ in element Ω_k is assumed to be a polynomial function. Considering the Legendre polynomials P up to degree $k = 1$ (corresponding to piecewise, linear functions) and a three-dimensional problem (i.e, $x_j = [x, y, z]$) yields

$$U_h(x_j,t)|_{\Omega_k}, v(x_j)|_{\Omega_k} \in P^1\left(\Omega_k\right), \tag{3.5}$$

where

$$P^1\left(\Omega_k\right) = \mathrm{span}\left\{\phi_0(x_j), \phi_1(x_j), \phi_2(x_j), \phi_3(x_j)\right\}$$
$$= \mathrm{span}\left\{1, x - x_{c_k}, y - y_{c_k}, z - z_{c_k}\right\}. \tag{3.6}$$

Here, the constants $[x_{c_k}, y_{c_k}, z_{c_k}]$ represent the centroid of the element Ω_k, defined by

$$x_{c_k} = \frac{\iiint_{\Omega_k} x\,dxdydz}{\iiint_{\Omega_k} dxdydz}, \quad y_{c_k} = \frac{\iiint_{\Omega_k} y\,dxdydz}{\iiint_{\Omega_k} dxdydz}, \quad z_{c_k} = \frac{\iiint_{\Omega_k} z\,dxdydz}{\iiint_{\Omega_k} dxdydz}. \tag{3.7}$$

The numerical solution variables (degrees of freedom) are then the cell average value $\overline{U}_k$, and the average value of the x, y, and z components of the solution gradient within each cell $[(\overline{\Delta_x U})_k,$

$(\overline{\Delta_y U})_k$, $(\overline{\Delta_z U})_k]$. The assumed solution in cell k is therefore

$$U_k = \overline{U}_k + (\overline{\Delta_x U})_k(x - x_{c_k}) + (\overline{\Delta_y U})_k(y - y_{c_k}) + (\overline{\Delta_z U})_k(z - z_{c_k}).$$ (3.8)

3.2.3   Update Formulas for the Degrees of Freedom

Inserting the solution representation from Equation (3.8) into the weak formulation shown in Equation (3.4) and performing integration using each test function yields the following update formulas for the degrees of freedom:

$$\overline{U}_k^{n+1} = \overline{U}_k^n - \frac{1}{V_k} \iint\limits_{\partial\Gamma_k \times T} F_i \cdot \hat{n}_i \, d\Gamma dt + \frac{1}{V_k} \iint\limits_{\Omega_k \times T} S \, dx_j dt,$$ (3.9)

$$\begin{bmatrix} \overline{\Delta_x U}_k^{n+1} \\ \overline{\Delta_y U}_k^{n+1} \\ \overline{\Delta_z U}_k^{n+1} \end{bmatrix} = \begin{bmatrix} \overline{\Delta_x U}_k^n \\ \overline{\Delta_y U}_k^n \\ \overline{\Delta_z U}_k^n \end{bmatrix} + K_k \left(- \iint\limits_{\partial\Gamma_k \times T} \begin{bmatrix} x - x_{c_k} \\ y - y_{c_k} \\ z - z_{c_k} \end{bmatrix} F_i \cdot \hat{n}_i \, d\Gamma dt \right.$$

$$+ \iint\limits_{\Omega_k \times T} \begin{bmatrix} F_x \\ F_y \\ F_z \end{bmatrix} dx_j dt$$ (3.10)

$$\left. + \iint\limits_{\Omega_k \times T} \begin{bmatrix} x - x_{c_k} \\ y - y_{c_k} \\ z - z_{c_k} \end{bmatrix} S \, dx_j dt \right).$$

Here, V_k is the volume of the cell, defined as

$$V_k = \iiint\limits_{\Omega_k} dx dy dz,$$ (3.11)

$\partial\Gamma_k$ represents the edges along the boundary Γ_k, and K_k is the matrix given by

$$K_k = \begin{bmatrix} I_{xx} & I_{xy} & I_{xz} \\ I_{yx} & I_{yy} & I_{yz} \\ I_{zx} & I_{zy} & I_{zz} \end{bmatrix}^{-1},$$ (3.12)

where

$$I_{xx} = \iiint_{\Omega_k} (x - x_{c_k})^2 \, dx_j, \tag{3.13a}$$

$$I_{xy} = I_{yx} = \iiint_{\Omega_k} (x - x_{c_k})(y - y_{c_k}) \, dx_j, \tag{3.13b}$$

$$I_{xz} = I_{zx} = \iiint_{\Omega_k} (x - x_{c_k})(z - z_{c_k}) \, dx_j, \tag{3.13c}$$

$$I_{yy} = \iiint_{\Omega_k} (y - y_{c_k})^2 \, dx_j, \tag{3.13d}$$

$$I_{yz} = I_{zy} = \iiint_{\Omega_k} (y - y_{c_k})(z - z_{c_k}) \, dx_j, \tag{3.13e}$$

$$I_{zz} = \iiint_{\Omega_k} (z - z_{c_k})^2 \, dx_j, \tag{3.13f}$$

$$\tag{3.13g}$$

are the area moments of inertia of the cell. Numerical approximations are now required for the evaluation of the interface flux, in addition to both the surface integral of the flux as well as the volume integral of the flux and source term.

3.2.4 Interface-Flux Approximation and Surface Integral

Solutions between cells are discontinuous in a DG scheme. In order to obtain the interface flux along the edge $\partial\Gamma_k$, a numerical Riemann solver is used to approximate

$$\boldsymbol{F}_i \cdot \hat{n}_i \approx \widetilde{\boldsymbol{F}}_\xi, \tag{3.14}$$

where $\widetilde{\boldsymbol{F}}_\xi$ is the vector of computed fluxes normal to the edge $\partial\Gamma_k$. Two examples of popular Riemann solvers (or flux functions, as they are commonly referred to) are Roe's flux function [35] and the Harten-Lax-Leer-Einfeldt (HLLE) flux function [16, 13]. Non-linear fluxes require that the integral be estimated by a quadrature rule. Cockburn and Shu proved that for a semi-discrete method with a solution representation in polynomial space P^k, a quadrature rule used for the integration in space along edges must be exact for a polynomial of at least degree $2k + 1$ [9]. Therefore, to achieve third-order accuracy, a two-point Gaussian quadrature is used for approximating the spatial integration. The quadrature rule is then

$$\iint_{\partial\Gamma_k \times T} \boldsymbol{F}_i \cdot \hat{n}_i \, d\Gamma dt \approx \int_T \sum_\xi w_\xi \widetilde{\boldsymbol{F}}_\xi \, dt, \tag{3.15}$$

where w_ξ is the weight of quadrature point ξ. For integration in time, the midpoint rule is used, thereby evaluating fluxes at $t^{n+\frac{1}{2}}$,

$$\iint\limits_{\partial\Gamma_k\times T} \boldsymbol{F}_i \cdot \hat{n}_i \, d\Gamma dt \approx \Delta t \sum_\xi w_\xi \widetilde{\boldsymbol{F}}_\xi^{n+\frac{1}{2}}. \tag{3.16}$$

These fluxes are used once more for approximating the surface integral in the slope update Equation (3.10), such that

$$\iint\limits_{\partial\Gamma_k\times T} \begin{bmatrix} x - x_{c_k} \\ y - y_{c_k} \\ z - z_{c_k} \end{bmatrix} \boldsymbol{F}_i \cdot \hat{n}_i \, d\Gamma dt \approx \Delta t \begin{bmatrix} \sum_\xi w_\xi \left(x_\xi - x_{c_k}\right) \widetilde{\boldsymbol{F}}_\xi^{n+\frac{1}{2}} \\ \sum_\xi w_\xi \left(y_\xi - y_{c_k}\right) \widetilde{\boldsymbol{F}}_\xi^{n+\frac{1}{2}} \\ \sum_\xi w_\xi \left(z_\xi - z_{c_k}\right) \widetilde{\boldsymbol{F}}_\xi^{n+\frac{1}{2}} \end{bmatrix}. \tag{3.17}$$

Suzuki chose to use the Radau IIA method for source-term time integration, adding a required state at time $t^{n+\frac{1}{3}}$ [38]. Thus, each quadrilateral element in three dimensions requires 16 quadrature points associated with times $t^{n+\frac{1}{6}}$, $t^{n+\frac{1}{2}}$ for the calculation of states $U_h(x_i, t^{n+\frac{1}{3}})$, $U_h(x_i, t^{n+1})$. A schematic of the surface integral with all necessary quadrature points is shown in Figure 3.1. This image is presented in two spatial dimensions for ease of illustration. The (x, y, z) coordinates of Gauss points $\boldsymbol{x}_{q_1}$, $\boldsymbol{x}_{q_2}$ on the edge of a cell are found with

$$\boldsymbol{x}_{q_1} = \boldsymbol{x}_m - \frac{\ell}{2\sqrt{3}}, \tag{3.18a}$$

$$\boldsymbol{x}_{q_2} = \boldsymbol{x}_m + \frac{\ell}{2\sqrt{3}}, \tag{3.18b}$$

where $\boldsymbol{x}_m$ is the midpoint of the edge, and ℓ is the vector containing lengths of the edge.

3.2.5 Hancock's Predictor Step

In order to complete the approximation of the surface integral, the solution at the half-time steps $U_h(x_i, t^{n+\frac{1}{2}})$, $U_h(x_i, t^{n+\frac{1}{6}})$ for each update are needed as an input to the approximate Riemann solver. In order to obtain values at these points, Hancock's implicit predictor step is used. This technique is illustrated in Figure 3.2; at each interface $x_{k\pm\frac{1}{2}}$, three characteristic lines are drawn (using the one-dimensional Euler equations as an example). Paying attention to the domain encompassed within these interfaces, it is realized that wave interactions with adjacent cells can be ignored up to the half time step $t^{n+\frac{1}{2}}$. The update formula for the cell-average value shown in Equation (3.9) can now be altered slightly by removing element-face interactions, yielding

$$\overline{U}_k^{n+\psi} = \overline{U}_k^n - \frac{1}{V_k} \iint\limits_{\partial\Gamma_k\times T'} \hat{\boldsymbol{F}}_i \cdot \hat{n}_i \, d\Gamma dt + \frac{1}{V_k} \iint\limits_{\Omega_k\times T'} \boldsymbol{S} \, dx_j dt. \tag{3.19}$$

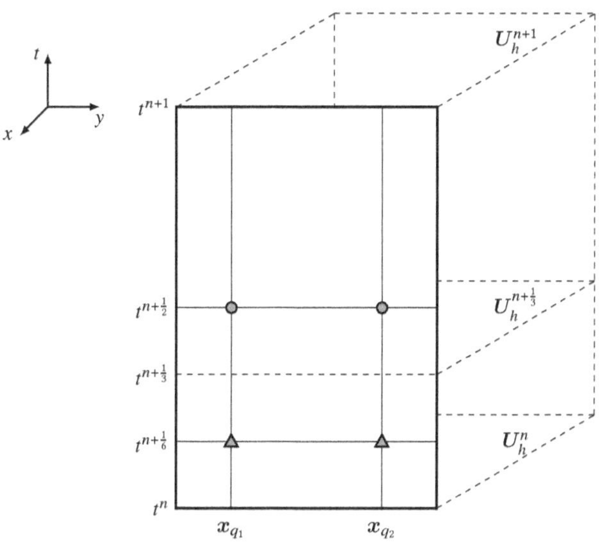

Figure 3.1: Schematic detailing the surface integral performed using two-point Gaussian quadrature in space and the midpoint rule in time. Quadrature points used for integration over the time step $[t^n, t^{n+\frac{1}{3}}]$ are represented as triangles ($\blacktriangle$), while bullets ($\bullet$) designate quadrature points used for integration over the time step $[t^n, t^{n+1}]$. At each quadrature point, a Riemann solver is used to compute an interface flux. Calculated states are designated with dashed lines.

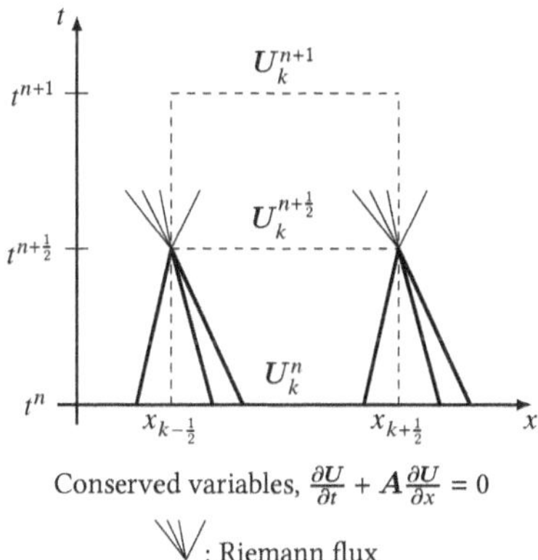

Figure 3.2: Hancock's technique for predicting values at a half time step $t^{n+\frac{1}{2}}$. Values here can be obtained solely based on information stored within the cell. These values are then used as inputs to the Riemann problems at $(x_{k\pm\frac{1}{2}}, t^{n+\frac{1}{2}})$.

Here, $\hat{F}_i$ is the flux vector evaluated using only the quadrature points and interior solution within element Ω_k, and T' is the time interval $\left[t^n, t^{n+\psi}\right]$ with $\psi = \frac{1}{2}$ or $\psi = \frac{1}{6}$. No approximate Riemann solver is needed for these fluxes, as they are evaluated directly using the flux dyad of the system. This predictor step can be simplified further through the use of the implicit Euler method to approximate the temporal integration of the source integral, thus evaluating the source at time $t^{n+\frac{1}{2}}$. The flux integral is evaluated at time t^n, and Equation (3.19) can then be written as

$$\overline{U}_k^{n+\psi} = \overline{U}_k^n + \frac{\Delta t}{V_k}\left[-\int_{\partial\Gamma_k} \hat{F}_i\left(U_h\left(x_j, t^n\right)\right) \cdot \hat{n}_i \, d\Gamma + \int_{\Omega_k} S\left(U_h\left(x_j, t^{n+\frac{1}{2}}\right)\right) dx_j\right]. \tag{3.20}$$

Once the predicted cell average state at a fractional-time step $\overline{U}_k^{n+\psi}$ is computed, the solution value at any point along the edge $\partial\Gamma_k$ can be found with

$$U_h\left(x_j, t^{n+\psi}\right) = \overline{U}_k^{n+\psi} + \phi_k^n \begin{bmatrix} \overline{\Delta_x U}_k^n \\ \overline{\Delta_y U}_k^n \\ \overline{\Delta_z U}_k^n \end{bmatrix} \cdot \begin{bmatrix} x - x_{c_k} \\ y - y_{c_k} \\ z - z_{c_k} \end{bmatrix}. \tag{3.21}$$

where ϕ_k^n is a vector of slope limiter values. Some options for slope limiting in multiple dimensions include the Venkatakrishnan limiter [40], which is the multi-dimensional extension to the one dimensional Van Albada limiter [1], or the Barth-Jesperson limiter [2], which is the multi-dimensional extension to the minmod slope limiter [23]. High-order finite-element methods (such as the DGH scheme) that rely on slopes for their assumed solution require a slope limiter to ensure monotonicity, as well as to prevent spurious oscillations in regions of high gradients.

The slope variables used here are the values at time t^n. To evaluate the spatial integration of the source term, one Gaussian quadrature point at the centroid of the element $(x_{c_k}, t^{n+\frac{1}{2}})$ is used. This integral then becomes

$$\int_{\Omega_k} S \, dx_j \approx V_k S\left(U_h\left(x_{c_k}, t^{n+\psi}\right)\right). \tag{3.22}$$

After inserting this definition into Equation (3.20), the state at a fractional time step $U_h(x_i, t^{n+\psi})$ can be found as

$$U_h(x_i, t^{n+\psi}) = \overline{U}_k^n + \phi_k^n \begin{bmatrix} \overline{\Delta_x U}_k^n \\ \overline{\Delta_y U}_k^n \\ \overline{\Delta_z U}_k^n \end{bmatrix} \cdot \begin{bmatrix} x - x_{c_k} \\ y - y_{c_k} \\ z - z_{c_k} \end{bmatrix}$$

$$- \frac{\Delta t}{V_k} \sum_{\partial\Gamma_k \in \partial\Omega_k} \int_{\partial\Gamma_k} \hat{F}_i\left(U_h\left(x_j, t^n\right)\right) \cdot \hat{n}_i \, d\Gamma + \Delta t S\left(U_h\left(x_{c_k}, t^{n+\psi}\right)\right). \tag{3.23}$$

The Hancock predictor step is performed at the beginning of each time step, with the solution at the half-time steps stored for later computations.

3.2.6 Volume Integral of Flux and Source Term

There are three volume integrals present in the update formulas for the degrees of freedom that must be evaluated; these are the source term, the flux, and the moment of the source term. The Radau IIA method for source-term time integration is used, due to the stiffness of the source term. This method is an L-stable, two-step, third-order accurate implicit time-marching scheme written as

$$y^{n+\frac{1}{3}} = y^n + \Delta t \left[\frac{5}{12} \left(\frac{dy}{dt} \right)^{n+\frac{1}{3}} - \frac{1}{12} \left(\frac{dy}{dt} \right)^{n+1} \right],$$

$$y^{n+1} = y^n + \Delta t \left[\frac{3}{4} \left(\frac{dy}{dt} \right)^{n+\frac{1}{3}} + \frac{1}{4} \left(\frac{dy}{dt} \right)^{n+1} \right]. \tag{3.24}$$

For spatial integration, Gaussian quadrature is used. The quadrature points used for the volume integrals are shown in Figure 3.3. Once again, for the ease of illustration, this image is presented in two spatial dimensions, with a four-point quadrature rule. For a three-dimensional reference element in the domain $[\zeta, \eta, \mu] \in [-1, 1]^3$, an eight-point quadrature rule is required. The position of these quadrature points are

$$\zeta_0 = \left(-\frac{1}{\sqrt{3}}, -\frac{1}{\sqrt{3}}, \frac{1}{\sqrt{3}} \right), \quad \zeta_1 = \left(\frac{1}{\sqrt{3}}, -\frac{1}{\sqrt{3}}, \frac{1}{\sqrt{3}} \right), \quad \zeta_2 = \left(\frac{1}{\sqrt{3}}, \frac{1}{\sqrt{3}}, \frac{1}{\sqrt{3}} \right),$$

$$\zeta_3 = \left(-\frac{1}{\sqrt{3}}, \frac{1}{\sqrt{3}}, \frac{1}{\sqrt{3}} \right), \quad \zeta_4 = \left(\frac{1}{\sqrt{3}}, -\frac{1}{\sqrt{3}}, -\frac{1}{\sqrt{3}} \right), \quad \zeta_5 = \left(-\frac{1}{\sqrt{3}}, -\frac{1}{\sqrt{3}}, -\frac{1}{\sqrt{3}} \right), \tag{3.25}$$

$$\zeta_6 = \left(-\frac{1}{\sqrt{3}}, \frac{1}{\sqrt{3}}, -\frac{1}{\sqrt{3}} \right), \quad \zeta_7 = \left(\frac{1}{\sqrt{3}}, \frac{1}{\sqrt{3}}, -\frac{1}{\sqrt{3}} \right).$$

This is the three-dimensional result of the Cartesian product for a one-dimensional Gaussian quadrature rule. Since elements can be of any polygonal shape, a mapping from the physical domain $[x, y, z]$ to a computational domain $[\zeta, \eta, \mu]$ is required. A trilinear mapping is chosen for this transformation, such that

$$x(\zeta, \eta, \mu) = \frac{1-\zeta}{2} \frac{1-\eta}{2} \frac{1+\mu}{2} x_0 + \frac{1+\zeta}{2} \frac{1-\eta}{2} \frac{1+\mu}{2} x_1 + \frac{1+\zeta}{2} \frac{1+\eta}{2} \frac{1+\mu}{2} x_2$$

$$+ \frac{1-\zeta}{2} \frac{1+\eta}{2} \frac{1+\mu}{2} x_3 + \frac{1+\zeta}{2} \frac{1-\eta}{2} \frac{1-\mu}{2} x_4 + \frac{1-\zeta}{2} \frac{1-\eta}{2} \frac{1-\mu}{2} x_5 \tag{3.26}$$

$$+ \frac{1-\zeta}{2} \frac{1+\eta}{2} \frac{1-\mu}{2} x_6 + \frac{1+\zeta}{2} \frac{1+\eta}{2} \frac{1-\mu}{2} x_7.$$

Here, x_i with $i = [0, 1, 2, \ldots, 7]$ are the physical positions of the nodes that make up a single element. The matrix of the coordinate transform from the physical domain to the computational domain is defined as the Jacobian, $J = \frac{\partial x_i}{\partial \zeta_j}$,

$$J(\zeta_j) = \frac{\partial x}{\partial \zeta} \frac{\partial y}{\partial \eta} \frac{\partial z}{\partial \mu} + \frac{\partial x}{\partial \eta} \frac{\partial y}{\partial \mu} \frac{\partial z}{\partial \zeta} + \frac{\partial x}{\partial \mu} \frac{\partial y}{\partial \zeta} \frac{\partial z}{\partial \eta} - \frac{\partial x}{\partial \mu} \frac{\partial y}{\partial \eta} \frac{\partial z}{\partial \zeta} - \frac{\partial x}{\partial \zeta} \frac{\partial y}{\partial \mu} \frac{\partial z}{\partial \eta} - \frac{\partial x}{\partial \eta} \frac{\partial y}{\partial \zeta} \frac{\partial z}{\partial \mu}. \tag{3.27}$$

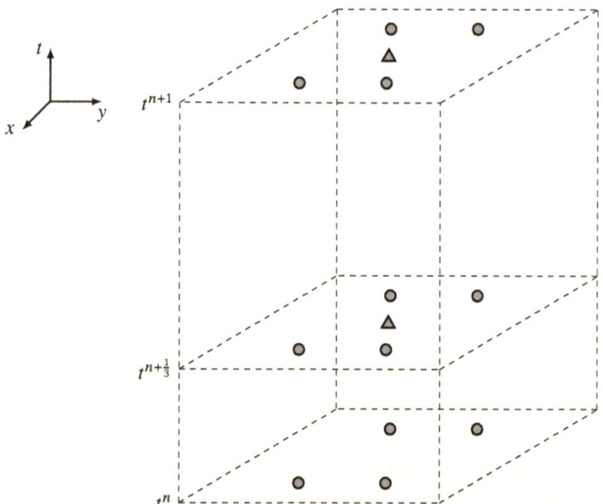

Figure 3.3: Schematic detailing volume integrals performed using Gaussian quadrature in space and Radau IIA in time. Quadrature points used for the source-term integration are presented as triangles (▲), while points used for flux integration are designated with bullets (•). Calculated states in time are shown with dashed lines.

Volume Integral of the Flux

The eight nodes of the trilinear mapping are used as quadrature points to approximate the volume integral, such that

$$\int_{\Omega_k} F_i\left(U(x_j,t)\right)\,\mathrm{d}x_j = \int_{\hat{\Omega}_k} F_i\left(U(\zeta_j,t)\right)\left|J(\zeta_j)\right|\,\mathrm{d}\zeta_j \approx \sum_{\chi=0}^{7} w_\chi |J(\zeta_j)| F_i\left(U(\zeta_j,t)\right), \qquad (3.28)$$

where χ is the index through the eight-point flux-volume quadrature with w_χ as the corresponding weight, and $|J(\zeta_j)|$ is the determinant of the Jacobian. Volume integrals for the flux appear in the update formula for the slope quantities. The temporal integration is again approximated by the two-point Radau IIA quadrature, such that

$$\iint_{\Omega_k \times T} F_i\,\mathrm{d}x_j\mathrm{d}t \approx \Delta t \int_{\Omega_k} \left[\frac{3}{4}F_i^{n+\frac{1}{3}} + \frac{1}{4}F_i^{n+1}\right]\,\mathrm{d}x_j. \qquad (3.29)$$

In space, the volume integral is replaced by Gaussian quadrature as mentioned,

$$\iint_{\Omega_k \times T} F_i\,\mathrm{d}x_j\mathrm{d}t \approx \Delta t \sum_{\chi} w_\chi |J(\zeta_j)| \left[\frac{3}{4}F_i^{n+\frac{1}{3}} + \frac{1}{4}F_i^{n+1}\right]. \qquad (3.30)$$

The volume integral of the flux for the time interval T' over $[t^n, t^{n+\frac{1}{3}}]$ is given by

$$\iint_{\Omega_k \times T'} F_i \, dx_j \, dt \approx \frac{\Delta t}{3} \sum_\chi w_\chi |J(\zeta_j)| \left[\frac{1}{2} F_i^n + \frac{1}{2} F_i^{n+\frac{1}{3}} \right]. \tag{3.31}$$

Volume Integral of the Source Term

The volume integral of the source term appears in the update formula for the cell-averaged values. Quadrature points used for approximating this integral are shown in Figure 3.3. A linearization of the source term is performed in order to make the update equation for the cell-averaged values independent to the other update formulas, such that

$$S(U_h(x_i, t)) \approx S(\overline{U}_k(t)) + \left(\frac{\partial S}{\partial U} \right)^n \left(\begin{bmatrix} \overline{\Delta_x U}_k^n \\ \overline{\Delta_y U}_k^n \\ \overline{\Delta_z U}_k^n \end{bmatrix} \cdot \begin{bmatrix} x - x_{c_k} \\ y - y_{c_k} \\ z - z_{c_k} \end{bmatrix} \right). \tag{3.32}$$

Integration in space for the source term is evaluated using a single quadrature point at the cell center, and Radau IIA is used once again to simplify the integral further, yielding

$$\iint_{\Omega_k \times T} S \, dx_j \, dt \approx V_k \Delta t \left[\frac{3}{4} S(\overline{U}_k^{n+\frac{1}{3}}) + \frac{1}{4} S(\overline{U}_k^{n+1}) \right]. \tag{3.33}$$

Volume Integral of the Moment of the Source Term

The volume integral of the moment of the source term appears in the update formula for the slope quantities. The same linearization assumed for the source-term volume integral is used here. The spatial integration is also done analytically, and the integration in time is approximated with Radau IIA quadrature, such that

$$\iint_{\Omega_k \times T} (x - x_{c_k}) S \, dx_j \, dt \approx \frac{3\Delta t}{4} \frac{\partial S}{\partial U}^{n+\frac{1}{3}} \left(I_{xx} \overline{\Delta_x U}_k^{n+\frac{1}{3}} + I_{xy} \overline{\Delta_y U}_k^{n+\frac{1}{3}} + I_{xz} \overline{\Delta_z U}_k^{n+\frac{1}{3}} \right)$$
$$+ \frac{\Delta t}{4} \frac{\partial S}{\partial U}^{n+1} \left(I_{xx} \overline{\Delta_x U}_k^{n+1} + I_{xy} \overline{\Delta_y U}_k^{n+1} + I_{xz} \overline{\Delta_z U}_k^{n+1} \right), \tag{3.34a}$$

$$\iint_{\Omega_k \times T} (y - y_{c_k}) S \, dx_j \, dt \approx \frac{3\Delta t}{4} \frac{\partial S}{\partial U}^{n+\frac{1}{3}} \left(I_{yx} \overline{\Delta_x U}_k^{n+\frac{1}{3}} + I_{yy} \overline{\Delta_y U}_k^{n+\frac{1}{3}} + I_{yz} \overline{\Delta_z U}_k^{n+\frac{1}{3}} \right)$$
$$+ \frac{\Delta t}{4} \frac{\partial S}{\partial U}^{n+1} \left(I_{yx} \overline{\Delta_x U}_k^{n+1} + I_{yy} \overline{\Delta_y U}_k^{n+1} + I_{yz} \overline{\Delta_z U}_k^{n+1} \right), \tag{3.34b}$$

$$\iint_{\Omega_k \times T} (z - z_{c_k}) S \, dx_j \, dt \approx \frac{3\Delta t}{4} \frac{\partial S}{\partial U}^{n+\frac{1}{3}} \left(I_{zx} \overline{\Delta_x U}_k^{n+\frac{1}{3}} + I_{zy} \overline{\Delta_y U}_k^{n+\frac{1}{3}} + I_{zz} \overline{\Delta_z U}_k^{n+\frac{1}{3}} \right)$$
$$+ \frac{\Delta t}{4} \frac{\partial S}{\partial U}^{n+1} \left(I_{zx} \overline{\Delta_x U}_k^{n+1} + I_{zy} \overline{\Delta_y U}_k^{n+1} + I_{zz} \overline{\Delta_z U}_k^{n+1} \right). \tag{3.34c}$$

3.2.7 Discrete Update Formulas

The discrete form of the cell-average update is then

$$
\begin{bmatrix} \overline{U}_k^{n+\frac{1}{3}} \\ \overline{U}_k^{n+1} \end{bmatrix} = \begin{bmatrix} \overline{U}_k^{n} \\ \overline{U}_k^{n} \end{bmatrix} - \frac{\Delta t}{V_k} \begin{bmatrix} \frac{1}{3} \sum_{\xi} w_\xi \widetilde{\boldsymbol{F}}^{n+\frac{1}{6}} \\ \sum_{\xi} w_\xi \widetilde{\boldsymbol{F}}^{n+\frac{1}{2}} \end{bmatrix} + \Delta t \begin{bmatrix} \frac{5}{12}\boldsymbol{I} & -\frac{1}{12}\boldsymbol{I} \\ \frac{3}{4}\boldsymbol{I} & \frac{1}{4}\boldsymbol{I} \end{bmatrix} \begin{bmatrix} \overline{\boldsymbol{S}}^{n+\frac{1}{3}} \\ \overline{\boldsymbol{S}}^{n+1} \end{bmatrix}.
\tag{3.35}
$$

Though some PDEs will contain source terms that can produce closed form solutions to Equation (3.35), Suzuki advocates using Newton's method for complex, non-linear source terms. The discrete form of the slope updates is written as

$$
\begin{bmatrix}
\overline{\Delta_x U}_k^{n+\frac{1}{3}} \\
\overline{\Delta_y U}_k^{n+\frac{1}{3}} \\
\overline{\Delta_z U}_k^{n+\frac{1}{3}} \\
\overline{\Delta_x U}_k^{n+1} \\
\overline{\Delta_y U}_k^{n+1} \\
\overline{\Delta_z U}_k^{n+1}
\end{bmatrix}
=
\begin{bmatrix}
\overline{\Delta_x U}_k^{n} \\
\overline{\Delta_y U}_k^{n} \\
\overline{\Delta_z U}_k^{n} \\
\overline{\Delta_x U}_k^{n} \\
\overline{\Delta_y U}_k^{n} \\
\overline{\Delta_z U}_k^{n}
\end{bmatrix}
+ \Delta t\, \Upsilon
\begin{bmatrix}
\frac{1}{3} \sum_\xi w_\xi (x_\xi - x_{c_k}) \widetilde{F}_\xi^{n+\frac{1}{6}} \\
\frac{1}{3} \sum_\xi w_\xi (y_\xi - y_{c_k}) \widetilde{F}_\xi^{n+\frac{1}{6}} \\
\frac{1}{3} \sum_\xi w_\xi (z_\xi - z_{c_k}) \widetilde{F}_\xi^{n+\frac{1}{6}} \\
\sum_\xi w_\xi (x_\xi - x_{c_k}) \widetilde{F}_\xi^{n+\frac{1}{2}} \\
\sum_\xi w_\xi (y_\xi - y_{c_k}) \widetilde{F}_\xi^{n+\frac{1}{2}} \\
\sum_\xi w_\xi (z_\xi - z_{c_k}) \widetilde{F}_\xi^{n+\frac{1}{2}}
\end{bmatrix}
$$

$$
+ \Delta t\, \Upsilon
\begin{bmatrix}
\frac{1}{3} \sum_\chi w_\chi |J(\zeta_j)| \left[\frac{1}{2}(F_x)_\chi^n + \frac{1}{2}(F_x)_\chi^{n+\frac{1}{3}} \right] \\
\frac{1}{3} \sum_\chi w_\chi |J(\zeta_j)| \left[\frac{1}{2}(F_y)_\chi^n + \frac{1}{2}(F_y)_\chi^{n+\frac{1}{3}} \right] \\
\frac{1}{3} \sum_\chi w_\chi |J(\zeta_j)| \left[\frac{1}{2}(F_z)_\chi^n + \frac{1}{2}(F_z)_\chi^{n+\frac{1}{3}} \right] \\
\sum_\chi w_\chi |J(\zeta_j)| \left[\frac{3}{4}(F_x)_\chi^{n+\frac{1}{3}} + \frac{1}{4}(F_x)_\chi^{n+1} \right] \\
\sum_\chi w_\chi |J(\zeta_j)| \left[\frac{3}{4}(F_y)_\chi^{n+\frac{1}{3}} + \frac{1}{4}(F_y)_\chi^{n+1} \right] \\
\sum_\chi w_\chi |J(\zeta_j)| \left[\frac{3}{4}(F_z)_\chi^{n+\frac{1}{3}} + \frac{1}{4}(F_z)_\chi^{n+1} \right]
\end{bmatrix}
\tag{3.36}
$$

$$
+ \Delta t
\begin{bmatrix}
\frac{5}{12} I & -\frac{1}{12} I \\
\frac{3}{4} I & \frac{1}{4} I
\end{bmatrix}
\begin{bmatrix}
\frac{\partial S}{\partial U}^{n+\frac{1}{3}} \overline{\Delta_x U}_k^{n+\frac{1}{3}} \\
\frac{\partial S}{\partial U}^{n+\frac{1}{3}} \overline{\Delta_y U}_k^{n+\frac{1}{3}} \\
\frac{\partial S}{\partial U}^{n+\frac{1}{3}} \overline{\Delta_z U}_k^{n+\frac{1}{3}} \\
\frac{\partial S}{\partial U}^{n+1} \overline{\Delta_x U}_k^{n+1} \\
\frac{\partial S}{\partial U}^{n+1} \overline{\Delta_y U}_k^{n+1} \\
\frac{\partial S}{\partial U}^{n+1} \overline{\Delta_z U}_k^{n+1}
\end{bmatrix}.
$$

Here, I is the identity matrix $\in \mathbb{R}^{3\times3}$, and Υ is the block matrix given by

$$
\Upsilon =
\begin{bmatrix}
K_k & 0 & 0 \\
0 & K_k & 0 \\
0 & 0 & K_k
\end{bmatrix}.
\tag{3.37}
$$

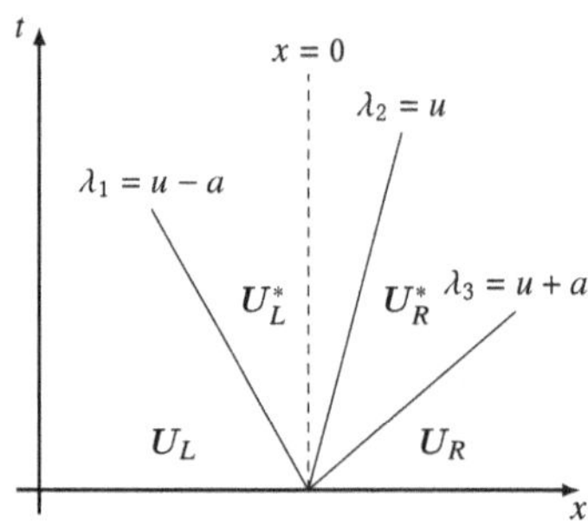

Figure 3.4: One-dimensional formulation of the Riemann problem.

3.3 Computation of Intercell Fluxes

As shown in Section 3.2.4, intercell fluxes on the boundary of computational elements must be evaluated for the DGH scheme. In this section, some concepts briefly mentioned earlier such as the Riemann problem and approximate Riemann solvers are described in more detail.

3.3.1 The Riemann Problem

The Riemann problem is an initial value problem for PDEs. For a one-dimensional system of conservation laws given by

$$\frac{\partial U}{\partial t} + \frac{\partial F}{\partial x} = 0,$$
(3.38)

the piecewise initial conditions for the Riemann problem can be expressed as

$$U(x, 0) = \begin{cases} U_L, & \text{if } x < 0 \\ U_R, & \text{if } x > 0 \end{cases},$$
(3.39)

where U_L and U_R are the left and right states of the initial discontinuity at $x = 0$ respectively. Figure 3.4 shows an example for the solution of the Riemann problem using the one-dimensional, hyperbolic Euler equations with eigenvalues, $\lambda_{1,2,3}$, and speed of sound, a. The states U_L^* and U_R^* represent the unknown regions in this problem. The Rankine-Hugonoit conditions can be used to solve for these states, and calculate the flux between U_L and U_R. For relatively simple sets of PDEs (such as the Euler equations), exact Riemann solvers (such as the one proposed by Gottlieb and Groth) can be used to determine this flux [14]. However, for more complicated systems, an exact solution cannot be found and approximate Riemann solvers must be used.

3.3.2 Approximate Riemann Solvers

Approximate Riemann solvers are used to solve the Riemann problem and evaluate the numerical flux between cells in a finite-element scheme.

Roe's Approximate Riemann Solver

For his approximate Riemann solver, Roe uses the fact that the solution to a linear system can be obtained in closed form through an eigensystem decomposition. The governing Equation (3.38) can be re-written as

$$\frac{\partial U}{\partial t} + A \frac{\partial U}{\partial x} = 0,$$ (3.40)

where the Jacobian matrix $A = \frac{\partial F}{\partial U}$. Roe wishes to build a matrix, $\bar{A}$, dependent on an intermediate state, $\hat{U}$, which itself is a function of the initial states U_L and U_R. In doing so, the non-linear conservation laws will have been replaced with a linearized system that is both stable and accurate.

There are a number of restrictions imposed on the Roe matrix $\bar{A}$ to ensure that it follows the behaviour of the original Jacobian:

- $\bar{A}$ must have both real eigenvalues and a complete set of linearly independent eigenvectors to ensure the new linear system retains hyperbolicity.

- In the limit that $U_L = U_R = U$, $\bar{A}$ should recover the same system as the original Jacobian, such that

$$\bar{A}(U_L, U_R) = \bar{A}(U) = \frac{\partial F}{\partial U}.$$ (3.41)

- For any given states, U_L and U_R, an exact solution exists such that

$$\Delta F = \bar{A} \Delta U.$$ (3.42)

The Roe-averaged state $\hat{U}$ is constructed through a corrected average approach demonstrated by Brown [8]. For example, when applied to the Gaussian closure, the primitive variables for this state are found with

$$\hat{\rho} = \sqrt{\rho_L \rho_R},$$ (3.43a)

$$\hat{u}_i = \frac{\sqrt{\rho_L} u_{i_L} + \sqrt{\rho_R} u_{i_R}}{\sqrt{\rho_L} + \sqrt{\rho_R}},$$ (3.43b)

$$\hat{W}_{ij} = \frac{\sqrt{\rho_L} W_{ij_L} + \sqrt{\rho_R} W_{ij_R}}{\sqrt{\rho_L} + \sqrt{\rho_R}} + \frac{1}{3} \frac{\sqrt{\rho_L \rho_R}}{\left(\sqrt{\rho_L} + \sqrt{\rho_R}\right)^2} \Delta u_i \Delta u_j,$$ (3.43c)

where $W_{ij} = \frac{P_{ij}}{\rho}$. The final form of the approximate flux function is then found as

$$F^* = \frac{1}{2}(F_L + F_R) - \frac{1}{2}|\bar{A}|(U_R - U_L).$$ (3.44)

Here, F_L, F_R are the left and right flux states of the interface. This form of the interface flux can be thought of as the average flux plus a dissipation term.

This linearization process of the Riemann problem is well suited for contact discontinuities and shock waves. However, the continuously changing state within a rarefaction wave is over-

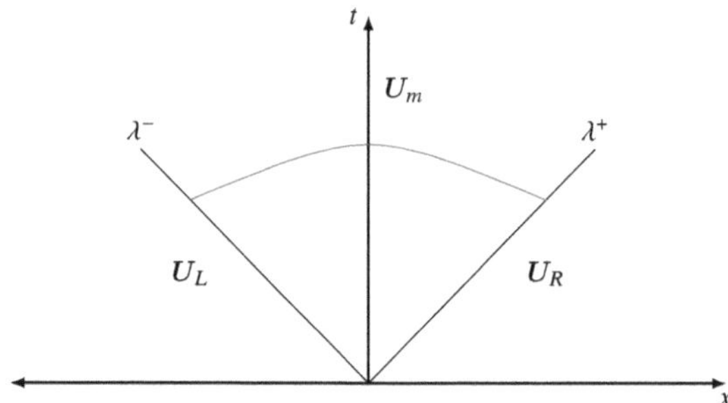

Figure 3.5: Illustration of the HLL approximate Riemann solver.

simplified through linearization, as the averaging procedure models all propagating waves using only their eigenvalues. As a result, the model is unable to distinguish between shock waves and rarefaction waves. This issue also leads to the carbuncle phenomenon. An entropy fix proposed by Harten [16] can be used as an extension to the approximate Roe solver in order to correct eigenvalues and avoid this problem. The positive and negative wavespeeds, $\lambda_k^\pm$, are replaced by a corrected value, $\hat{\lambda}_k^\pm$, such that

$$|\hat{\lambda}_k^\pm| = \begin{cases} |\lambda_k|, & \text{if } |\lambda_k| > \frac{\delta}{2} \\ \frac{|\lambda_k|^2}{\delta} + \frac{\delta}{4}, & \text{otherwise} \end{cases}, \tag{3.45}$$

where

$$\delta = \max\left[0, 4\left(\lambda_{k_R} - \lambda_{k_L}\right)\right]. \tag{3.46}$$

The HLL and HLLE Approximate Riemann Solvers

In the approximate Riemann solver proposed by Harten, Lax and van Leer [16], the one-dimensional conservation form shown in Equation (3.38) is integrated over space and time, such that

$$\iint_S \left[\frac{\partial U}{\partial t} + \frac{\partial F}{\partial x}\right] dx\, dt = 0, \tag{3.47}$$

where S is the space-time domain $[x_L, x_R] \times [0, \Delta t]$. Here, $x_L = \lambda^- \Delta t$, $x_R = \lambda^+ \Delta t$. Applying Green's theorem transforms this into an integral around the surface of the space-time domain, such that

$$\oint_S (U\, dx - F\, dt) = 0. \tag{3.48}$$

As shown in Figure 3.5, it is assumed that only two waves propagate in opposite directions with speeds λ^- and λ^+, with a single state U_m between them. These wavespeeds are the largest positive and negative speeds arising from the solution of the Riemann problem. Upon taking a time step of

Δt, the left and right waves will have moved a distance of $\lambda^-\Delta t$ and $\lambda^+\Delta t$ respectively. Evaluating the path integral in Equation (3.48) yields

$$-U_L\lambda^-\Delta t + U_R\lambda^+\Delta t - U_m\left(\lambda^+ - \lambda^-\right)\Delta t + F_L\Delta t - F_R\Delta t = 0. \tag{3.49}$$

Here, F_L and F_R are the flux vectors evaluated on the left and right side of the discontinuity. Rearranging for U_m gives

$$U_m = \frac{U_R\lambda^+ - U_L\lambda^- - (F_R - F_L)}{\lambda^+ - \lambda^-} \tag{3.50}$$

The goal of this approximate solver is to determine the flux at $x = 0$ or F_m. Simply evaluating the flux vector using state U_m such that $F_m = F(U_m)$ will not satisfy the Rankine-Hugonoit condition $\Delta F = \lambda\Delta U$. A second path integral is performed over the portion of the domain on the left or right side of the $x = 0$ discontinuity. If the right portion is chosen, this integral yields

$$U_R\lambda^+\Delta t - U_m\lambda^+\Delta t - F_R\Delta t + F_m\Delta t = 0. \tag{3.51}$$

Rearranging,

$$F_m = F_R + \lambda^+\left(U_m - U_R\right). \tag{3.52}$$

Substituting this back into Equation (3.50) gives

$$F_m = \frac{\lambda^+ F_L - \lambda^- F_R}{\lambda^+ - \lambda^-} + \frac{\lambda^-\lambda^+}{\lambda^+ - \lambda^-}\left(U_R - U_L\right). \tag{3.53}$$

The wavespeeds here for the HLL flux function are simply taken to be $\lambda^- = u_L - a_L$ and $\lambda^+ = u_R + a_R$. For the extension developed by Einfeldt [13], the wavespeeds are found by comparing velocities obtained from the Roe-averaged state and those of the left and right states. The HLLE approximate Riemann solver therefore requires the Roe-averaged state be known for the given system. Again taking the Euler equations as an example, Einfeldt proved stable schemes can be obtained with

$$\lambda^- = \min(u_L - a_L, \hat{u} - \hat{a}), \tag{3.54a}$$
$$\lambda^+ = \max(u_R + a_R, \hat{u} + \hat{a}), \tag{3.54b}$$

where $\hat{u}$ and $\hat{a}$ are the Roe-averaged speed and speed of sound obtained from $\hat{U}(U_L, U_R)$.

Chapter 4

Parallel Computing

4.1 Introduction

The discontinuous-Galerkin Hancock method was implemented within an academic framework intended for large-scale simulations of PDEs on modern distributed clusters. This framework uses adaptive mesh refinement and load balancing to achieve high efficiency while using up to hundreds of CPUs. During computations, the solution blocks that make up the mesh can be refined, coarsened, and distributed using Open Multi-Processing (OMP) and the Message Passing Interface (MPI) amongst different threads and processes. These structured blocks contain a number of nodes, which are described using i, j, and k indexing in three-dimensions. These blocks can be refined in any of these directions or any combination thereof, resulting in two or more child blocks. As refinement always produces blocks containing the same number of cells, it is assumed that the computational cost of all individual blocks is roughly equal.

4.2 Open Multi-Processing and the Message Passing Interface

OMP is an application programming interface (API) that supports shared-memory parallelization for applications through the use of library routines, compiler directives, and environment variables. Work on a local process can be forked and distributed amongst multiple threads, which execute parallelized sections of code. These threads share memory, but each has its own stack, used to track function calls.

MPI is a standard that defines library routines specific to message passing between processes on a cluster or network. When initialized, each MPI process is given a unique integer identifier referred to as its rank. This rank serves as a way to specify the destination and source of messages sent from process to process. Unlike OMP, MPI processes do not share memory.

OpenMP and MPI are currently used to distribute the work-load of parallel computations. In multi-block meshes, solution blocks are spread across available MPI processes on the network of large-scale clusters, and are further sub-divided amongst OpenMP threads. As adaptive mesh refinement is performed and the number of blocks in the mesh is altered, load-balancing is used in order to keep the number of blocks on each MPI process as even as possible.

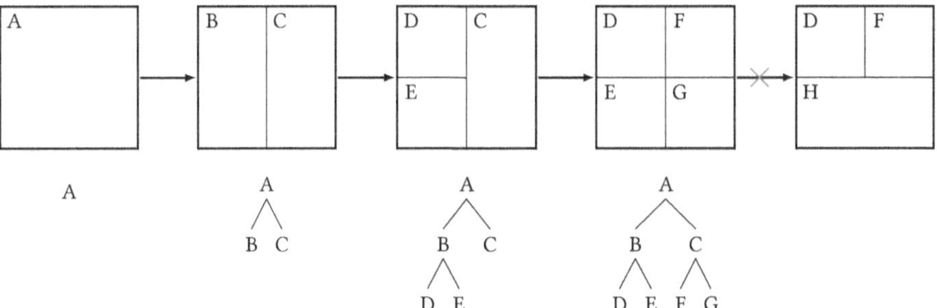

Figure 4.1: Block-based refinement anisotropic refinement, with the corresponding binary tree.

4.3   Adaptive Mesh Refinement

Adaptive mesh refinement is a technique widely used in numerical analysis for dynamically adjusting the accuracy of a solution during a simulation. Often, a sensitivity criterion or "refinement function" is used to algorithmically determine sections of low or high interest at a certain time within a computational domain. The resolution of these areas can then be either increased or decreased according to specified parameters of the refinement function. There are a few notable methods for mesh refinement that have become widely used, such as cell-based refinement, developed by Berger and Leveque [4]; patch-based refinement, developed by Berger and Collela [3]; and finally block-based refinement, originally presented by Quirk and Hanebutte [32], as well as Berger [5]. In the current implementation of the parallel framework, block-based refinement is used due to its simplicity and ease for load-balancing.

4.3.1   Block-Based Adaptive Mesh Refinement

By combining aspects of cell-based and patch-based refinement, block-based refinement yields a simple approach to AMR. For example, in two-dimensions, an initial block of $N \times N$ cells (considered to be the "parent" block) can be refined into multiple "child" blocks, each with $N \times N$ cells. These blocks can then be distributed on a parallel cluster to easily re-balance the local computational work load. Because high parallel efficiency is easy to obtain and it retains the advantages of structured meshes, block-based refinement is highly favoured. However, the connections between refined blocks as well as their refinement history has to be stored. A binary tree is typically used for this application [42]. Figure 4.1 illustrates a simple multi-stage case of block-based refinement. First, block A is refined into blocks B and C. Block B is then refined into blocks D and E, and finally block C is refined into blocks F and G. The accompanying binary tree at each stage is shown below the block representation. However, a problem arises when attempting to coarsen child blocks E and G into parent block H. This would be a valid mesh, but because D and F are on different branches of the refinement history tree, many implementations cannot allow them to be combined into a parent block. The only option for coarsening blocks together is to work

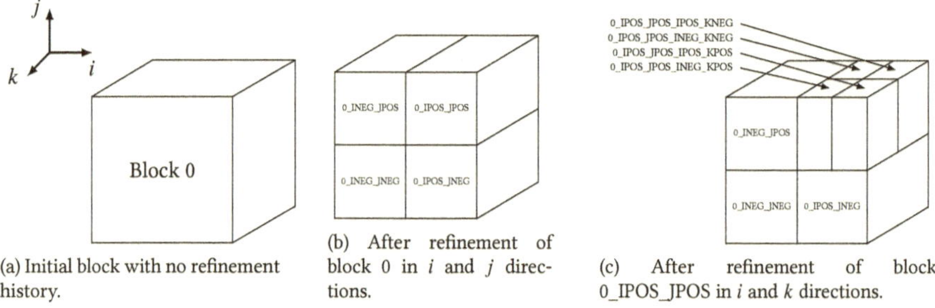

(a) Initial block with no refinement history.

(b) After refinement of block 0 in i and j directions.

(c) After refinement of block 0_IPOS_JPOS in i and k directions.

Figure 4.2: Block signature assignment with refinement-events.

backwards up the binary tree. In order to get around this limitation, a block-based adaptive mesh refinement scheme that does not use a tree to represent refinement history was implemented in the parallel framework.

4.4 Block Connectivity

As blocks are refined, coarsened, and possibly distributed across a network using MPI and OpenMP, it is necessary to retain information about neighbouring connections, as fluxes from block to block must be computed. In the current implementation, this is achieved through the use of block signatures, boundary signatures, and interblock faces.

4.4.1 Block Signature

Every block in the parallel framework is assigned a unique block signature. This consists of an arbitrary, unique root number, and a list of refinement events. An example of how block signatures are assigned with refinement is shown in Figure 4.2. As can be seen, when a refinement-event takes place and a block is refined, the child block inherits the parent's signature and additionally pushes back the new refinement event information. This information is comprised of the direction in which the block was refined, as well as whether the block was on the positive or negative side of the event per the local coordinate system. This signature serves as a unique identifier for a block, allowing for easy sorting of a list of blocks, as well as tracking a block's refinement history. Sorting of the block signatures is required to ensure all CPUs keep track of the blocks in a similar fashion.

When comparing two block signatures, the root number is first compared. Should they be equal, the first non-matching element in the refinement history of the blocks determines which signature comes first in a defined ordering. Lexicographical ordering is used to compare the refinement event information; for example, "INEG" will come before "IPOS", but "KNEG" will come after "JPOS".

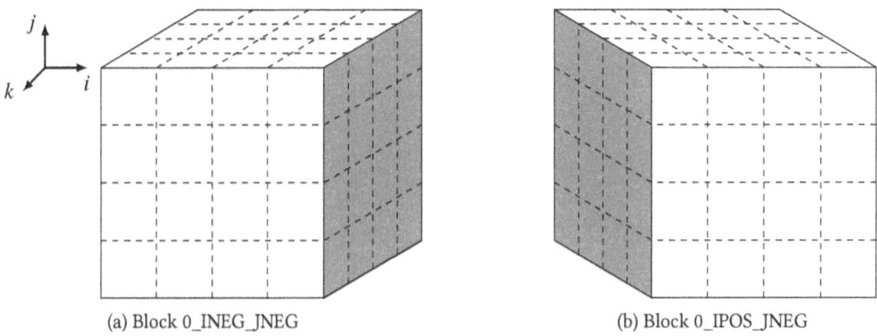

(a) Block 0_INEG_JNEG   (b) Block 0_IPOS_JNEG

Figure 4.3: Two neighbouring blocks, connected on the IPOS and INEG sides respectively.

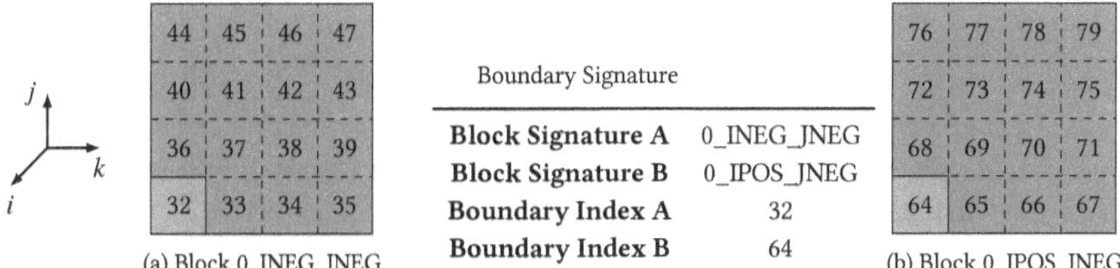

(a) Block 0_INEG_JNEG

Boundary Signature	
Block Signature A	0_INEG_JNEG
Block Signature B	0_IPOS_JNEG
Boundary Index A	32
Boundary Index B	64

(b) Block 0_IPOS_JNEG

Figure 4.4: Boundary faces with indices for the two blocks, with example boundary signature shown for connected faces highlighted in green.

4.4.2 Boundary Signature

As soon as there is more than one block representing the computational domain, interblock connectivity between blocks is required for the calculation of fluxes on the block boundaries. An example is shown in Figure 4.3. Boundary signatures contain the signatures of both connected blocks, as well as a unique index specifying the boundaries on either block. These boundaries are defined to have a side "A" and a side "B". Thus, each boundary signature is comprised of "A" and "B" side block signatures, as well as "A" and "B" cell indices. Every boundary signature is unique, due to the information stored. An example of a boundary signature for the connected blocks is shown in Figure 4.4. Similarly to block signatures, boundary signatures can be sorted. When comparing two boundary signatures, the signatures of block "A" are compared first (in lexicographical order, as previously described). Should they be equal, the boundary index of block "A" is then compared. If this is equal as well, the process is repeated using the information for block "B" in both signatures.

4.4.3 Composite Interblocks and Interblock Faces

As blocks are refined and coarsened, the connections between blocks on their boundaries have to be updated to reflect the changes made to the mesh. Often, these connections will be between

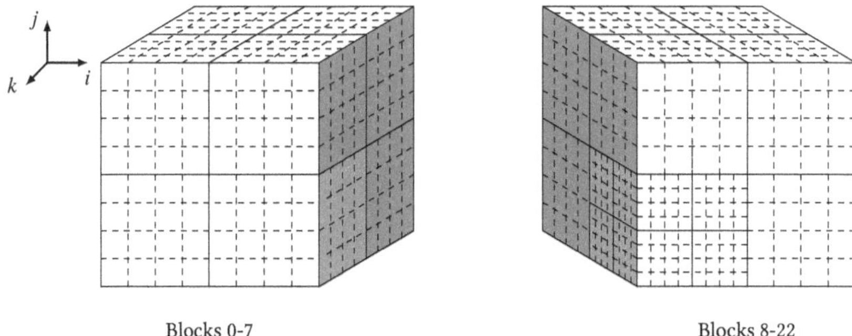

Blocks 0-7   Blocks 8-22

Figure 4.5: Interblock connection between blocks of differing refinement levels. Block connections highlighted in cyan indicate symmetric levels of refinement, while block connections in orange indicate a disparity.

blocks at different levels of refinement, with some boundaries on the coarser block having to communicate with multiple boundaries on the more refined side. An example of this is shown in Figure 4.5. In order to maintain these connections, two objects have been defined; interblock faces, and composite interblocks. Interblock faces represent a boundary face or a portion of a boundary face on a block. Each interblock face stores the nodes that constitute a mapping of the boundary face (in two-dimensions, two nodes are needed for a linear mapping, and in three-dimensions, four nodes for a bilinear mapping), the boundary signature for its connection, the CPU rank on which the neighbour block is stored, and several pointers. These pointers are handled externally and provide the interblock face with information about refinement events, CPU updates, as well as solution data for flux computations.

Composite interblocks are simply collections of these interblock faces that constitute the entire cell face. To be general, cardinal directions can be defined for the orientation of a block. The cardinal directions of a block represent the directions in which the face is defined (i, j, k), as well as whether this index increases or decreases. Going back to Figure 4.4, the first and second cardinal direction for the connected faces shown are both "KPOS" and "JPOS" respectively. However, it is possible to connect the boundary faces of blocks with differing cardinal directions. To keep track of the orientation of connected blocks, the cardinal direction(s) of both blocks as well as the side ("A" or "B") of the local block are also stored within the composite interblock. Once again referring to Figure 4.5, each of the block connections highlighted in orange require a composite interblock containing four interblock faces that constitute the boundary face due to the difference in refinement levels of the connected blocks, whereas the connections highlighted in cyan are composite interblocks containing a single interblock face that makes up the entire boundary face.

When adaptive mesh refinement is performed, refinement pointers contained within each interblock face are updated in order to inform the interblock about how the two connected blocks are changing. In case of refinement, the contained interblock face will split itself appropriately, and the composite interblock can also divide itself as needed. When coarsening, composite interblocks will recognize that some contained interblock faces may need to be re-combined, and will create parent interblock faces out of the children.

Gaussian quadrature points created by the mappings contained in the interblock face are used for computing fluxes on block boundaries. These are colloquially referred to as boundary flux quadrature points. Utilizing pointers to solution data, the states on either side of the quadrature point are used as inputs to a Riemann problem. This step requires solution data be sent across the network at every time step prior to computing boundary fluxes.

4.5   The Rainbow Connection

In order to manage message passing between blocks on different CPUs, each process has a single hub for interprocess communication, whimsically referred to as the rainbow connection. This object manages information buffers for data required to be sent from one MPI process to another, such as solution states on either side of a cell boundary for flux computations. Other data managed by the rainbow connection includes refinement-event information, as well as CPU ranks for load-balancing.

Whenever blocks are refined, coarsened, or moved from process to process, its interblock connections are registered with the rainbow connection by passing its boundary signature, as well as the CPU rank of the neighbouring block. After all blocks have registered their connections, they are sorted within the rainbow connection. Memory is then allocated for both send and receive buffers for solution data, refinement information, and CPU information. The interblock faces link their send and receive pointers to the appropriate segment of these allocated buffers by once again passing their boundary signature and neighbour CPU rank. The rainbow connection ensures that the send and receive pointers of two connected boundaries will point to the same locations in memory (but flipped for the appropriate block). A simple example of these connections is shown in Figure 4.6.

For maximum efficiency, MPI messages are sent as early as possible during each time step, with the largest bulk of local work done while waiting for all messages to be received across the network. With the DGH scheme, there are two local messages that must be sent per time step when a slope limiter is used, in addition to a globally communicated message for the time step value. When no slope limiter is used, there is only one message that must be communicated between neighbouring blocks. Due to the large amount of local work that can be done while these messages are being sent, it remains a highly efficient scheme.

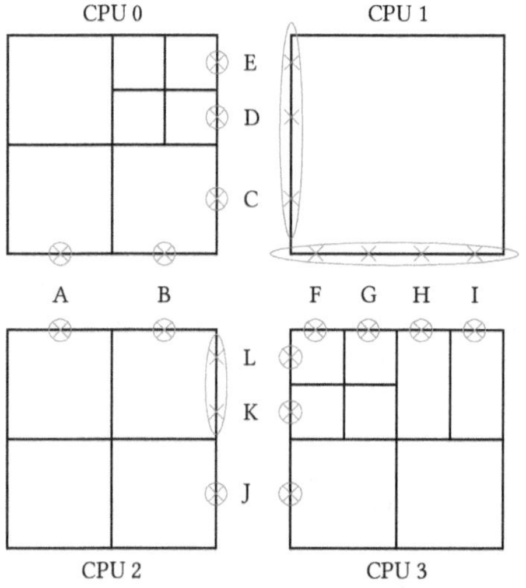

Rainbow connection buffers.

CPU	Neighbour CPU	Send/Receive Buffers
0	1	C, D, E
	2	A, B
1	0	C, D, E
	3	F, G, H, I
2	0	A, B
	3	J, K, L
3	1	F, G, H, I
	2	J, K, L

Figure 4.6: Rainbow connection representation for a number of composite interblock connections.

Chapter 5

Numerical Results

5.1 Introduction

In this chapter, a wide range of hyperbolic and hyperbolic-relaxation systems are solved in order to demonstrate the robustness of the DGH method, as well as the generality of the current implementation. The first case is presented in Section 5.2, in which the linear convection-relaxation equation is solved in order to prove the implementation of the scheme is indeed third-order accurate in up to three dimensions.

Next, in Section 5.3, the compressible Euler equations are solved in one, two and three dimensions. The one-dimensional cases shown are all classical problems which have been studied extensively in previous literature, and serve as strong proof for the accuracy of the DGH scheme. The two-dimensional Euler cases all demonstrate different aspects of the implementation; The solution of Ringleb's flow is the first proof that the DGH method is third-order on non-Cartesian meshes for a non-linear set of PDEs; other problems shown such as the double Mach reflection and bump case help demonstrate the efficacy of the AMR algorithm in two dimensions. One of the most impressive results is shown in the form of the Kelvin-Helmholtz instability, in which the low numerical dissipation of the scheme is demonstrated on a static mesh. A three-dimensional case is also presented to show the AMR algorithm's ability to track moving regions of high interest in three dimensions.

In Section 5.4, the shallow water equations are solved in two-dimensions to further demonstrate the versatility of the DGH scheme.

The Gaussian Closure is solved in two dimensions in Section 5.5 to validate the application of the scheme to models derived from the kinetic theory of gases. A classical viscous solution is computed and compared to a traditional Navier-Stokes result. An example of multi-phase flow is also simulated.

The fourteen moment closure is solved in Section 5.6, demonstrating the application of the DGH method to even higher-order members of the maximum-entropy closure hierarchy.

Finally, strong scaling studies are performed in Section 5.7 to show the efficacy of the DGH method for parallel computations on modern distributed clusters.

5.2 Three-Dimensional Linear Convection-Relaxation Equation

The linear convection-relaxation PDE can be written as

$$\frac{\partial \rho}{\partial t} + u_i \frac{\partial \rho}{\partial x_i} = -\frac{\rho}{\tau}.$$ (5.1)

In three dimensions, this expands to

$$\frac{\partial \rho}{\partial t} + u_x \frac{\partial \rho}{\partial x} + u_y \frac{\partial \rho}{\partial y} + u_z \frac{\partial \rho}{\partial z} = -\frac{\rho}{\tau},$$ (5.2)

and the exact solution to this PDE is known to be

$$\rho(x_i, t) = \rho_0(x_i - u_i t) \exp\left(-\frac{t}{\tau}\right).$$ (5.3)

For the following problem a grid on the bounds $(x, y, z) \in [-10, 10]^3$ is used, on which the initial condition is given by

$$\rho(x_i, 0) = \rho_0(x_i) = \exp\left(-0.5 \left[x^2 + y^2 + z^2\right]\right).$$ (5.4)

For this case, the velocity components are $u_x = u_y = u_z = -2$, and the relaxation time is set as $\tau = 1$. The exact solution for this problem would then be given by

$$\rho(x, y, z, t) = \exp\left(-0.5 \left[(x - u_x t)^2 + (y - u_y t)^2 + (z - u_z t)^2\right]\right) \exp\left(-\frac{t}{\tau}\right).$$ (5.5)

The solution is time marched to a final non-dimensional time of $t = 3$, and is shown in Figure 5.1. The ℓ^2-norm of the error was then computed as

$$\ell^2 \text{ Error} = \sqrt{\sum_{i=0}^{N} (\rho_{i_e} - \rho_i)^2 \times V_i},$$ (5.6)

where N is the total number of elements, and ρ_{i_e}, ρ_i, V_i are the values of the exact solution, computed solution, and cell volume respectively in cell i. The ℓ^2-norm of the error computed for an increasing number of elements N are shown in Table 5.1. The order of accuracy for a solution, O_j, is found with

$$O_j = \frac{\ln \frac{\text{Err}_{j-1}}{\text{Err}_j}}{\ln \frac{N_j}{N_{j-1}}}.$$ (5.7)

The expected third-order accuracy of the scheme is recovered with this study. No slope limiter was used for this problem, and the HLL flux function was used.

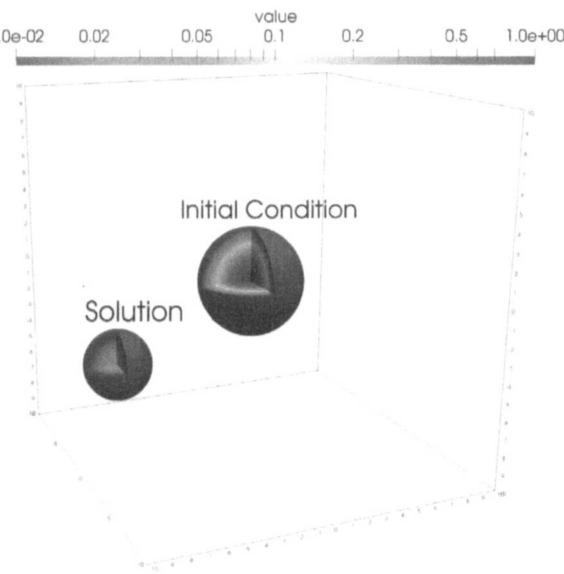

Figure 5.1: Initial condition and solution at $t = 3$ for linear convection-relaxation case in three dimensions.

Table 5.1: Convergence study for three-dimensional linear convection-relaxation case.

Number of Elements	ℓ^2 Error	Order
$80 \times 80 \times 80$	9.030×10^{-4}	—
$160 \times 160 \times 160$	1.231×10^{-4}	2.87
$200 \times 200 \times 200$	6.376×10^{-5}	2.95
$300 \times 300 \times 300$	1.911×10^{-5}	2.97
$350 \times 350 \times 350$	1.207×10^{-5}	2.98

5.3 The Euler Equations

The compressible Euler equations can be written in balance law form as

$$\frac{\partial U}{\partial t} + \frac{\partial F_i}{\partial x_i} = 0,$$ (5.8)

where

$$U = \begin{bmatrix} \rho \\ \rho u_i \\ \frac{p}{\gamma-1} + \frac{1}{2}\rho u_i u_i \end{bmatrix}, \quad F_i = \begin{bmatrix} \rho u_i \\ \rho u_i u_j + p\delta_{ij} \\ u_j\left(\frac{\gamma p}{\gamma-1} + \frac{1}{2}\rho u_i u_i\right) \end{bmatrix}.$$ (5.9)

The primitive solution vector W is given by

$$W = \begin{bmatrix} \rho \\ u_i \\ p \end{bmatrix}.$$ (5.10)

The Euler equations describe a gas in local thermodynamic equilibrium everywhere, and are the lowest-order member of the maximum-entropy moment hierarchy. The Venkatakrishnan slope limiter [40] was used for all the simulations performed in this section, unless otherwise mentioned.

5.3.1 One-Dimensional Problems

A number of classical Euler problems are solved in this section to demonstrate the use of the DGH method for the solution of one-dimensional problems. The problems shown here have been studied in a number of publications, and verifying the results obtained using the one-dimensional implementation of the solver with results shown previously in literature is easily done.

Shu Osher Shock

This problem first put forward by Shu and Osher [37] simulates a Mach 3 shock wave moving past a flow with a sinusoidal density fluctuation. Capturing the smooth oscillations as the shock propagates through the domain is the real challenge for this test case. The simulation is conducted on a one-dimensional grid with bounds $x \in [-5, 5]$, and the initial conditions in primitive form are given by

$$W_0 = \begin{cases} W_L, & x < -4 \\ W_R, & x > -4 \end{cases}, \quad W_L = \begin{bmatrix} 3.8571413 \\ 2.629369 \\ 10.333 \end{bmatrix}, \quad W_R = \begin{bmatrix} 1 + 0.2\sin(5x) \\ 0 \\ 1 \end{bmatrix}.$$ (5.11)

Zero-gradient boundary conditions are employed on both ends of the grid, and the solution is time marched to a final non-dimensional time of $t = 1.8$. The Roe flux function is used with a CFL number of 0.3, and solutions obtained on grids of 500 and 10000 elements are shown in Figure 5.2. The peaks and troughs of the density, Mach number and temperature fields aren't quite obtained with 500 cells, but the general structure of the solution remains smooth. With 10000 elements, the solution is in strong agreement with the results presented by Shu and Osher.

Blast Wave

This is a test case first used by Woodward & Colella [43]. The case generates many strong shocks, and is a rather difficult problem to solve accurately due to the numerous interactions involving rarefactions and contact discontinuities. The initial conditions are $\rho = 1$, $u = 0$ everywhere, and a piecewise pressure function such that

$$p(x) = \begin{cases} 1000, & 0 \leq x < 0.1, \\ 0.01, & 0.1 \leq x \leq 0.9, \\ 100, & 0.9 < x \leq 1.0. \end{cases} \tag{5.12}$$

The computation is performed on a one dimensional grid with bounds $x \in [0, 1]$, and both boundary conditions are reflecting walls. The solution is time marched to a final time of $t = 0.038$ with a CFL number of 0.3, using the Roe flux function. Once again, the solution obtained on a grid of 500 elements is compared to the solution obtained using a grid of 10000 elements in Figure 5.3. The solution is in great agreement with the results shown by Woodward and Colella. The solution obtained with 500 elements does not quite capture the peaks and troughs of the density and temperature field, but is still quite smooth and does a good job of capturing the general shape.

Noh Problem

This is the classical one-dimensional Noh problem described first in 1987 [30]. For this Riemann problem, a gas with a ratio of specific heats of $\gamma = 5/3$ is simulated on a domain $x \in [0, 1]$ with the initial conditions

$$W_0 = \begin{cases} W_L, & x < 0.5 \\ W_R, & x > 0.5 \end{cases}, \quad W_L = \begin{bmatrix} 1 \\ 1 \\ 1 \times 10^{-6} \end{bmatrix}, \quad W_R = \begin{bmatrix} 1 \\ -1 \\ 1 \times 10^{-6} \end{bmatrix}. \tag{5.13}$$

In this problem, two shocks of infinite strength move outwards from the initial discontinuity at $x_0 = 0.5$. Zero-gradient boundary conditions are used, and the solution is time marched to a final time of $t = 1$. Solutions obtained on grids of 500 and 5000 elements using a CFL number of 0.3 and the Roe flux function are shown in Figure 5.4. The solutions admit a slight entropy error at x_0 within the density and temperature plot, but the solutions are symmetric and do not contain any noticeably strong oscillations, which are both important factors for this test.

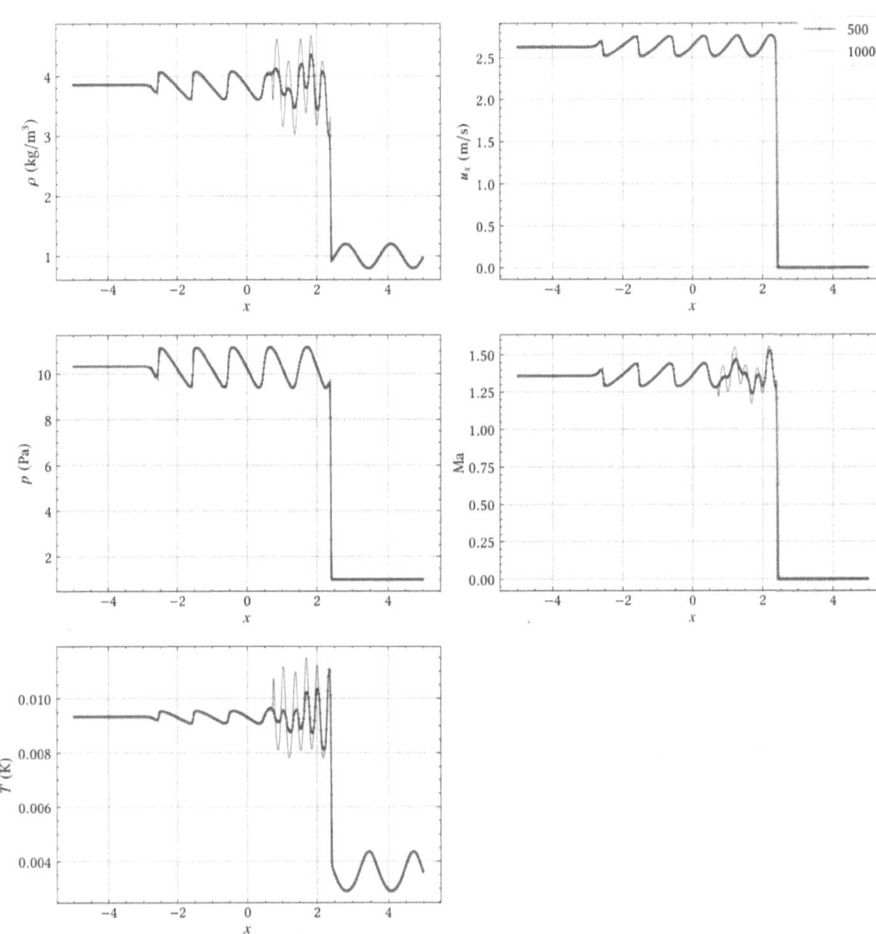

Figure 5.2: Shu Osher shock problem at a final time of $t = 1.8$, using the Roe flux function.

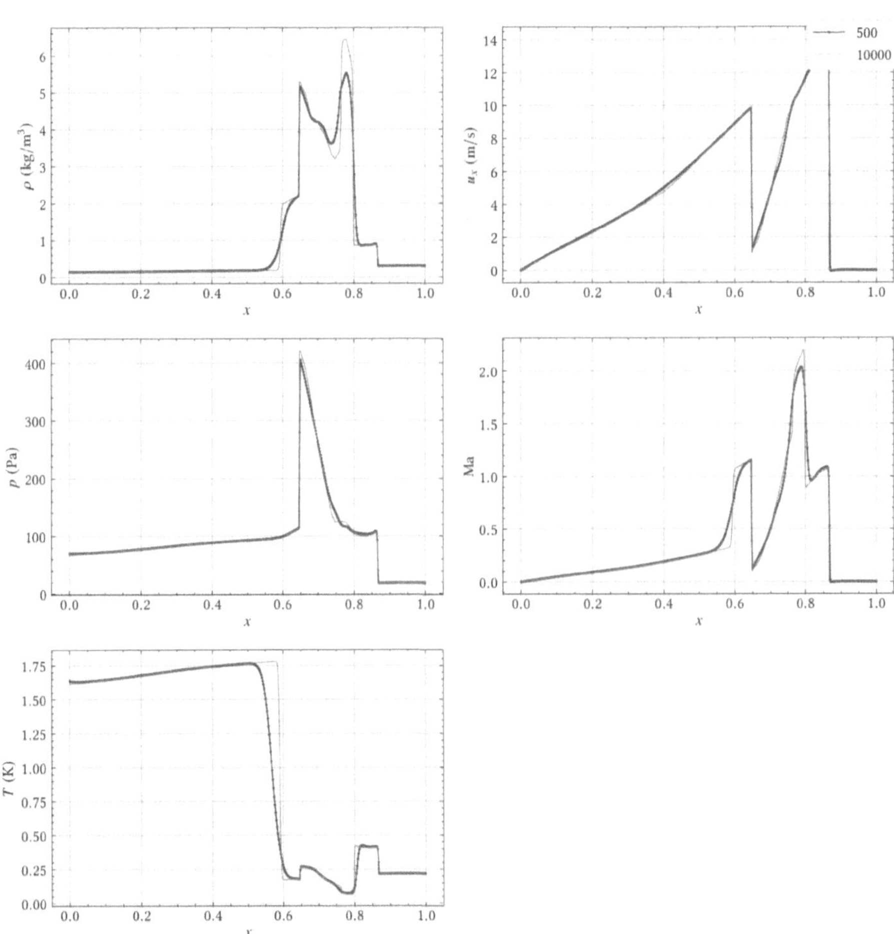

Figure 5.3: Blast wave problem at a final time of $t = 0.038$, using the Roe flux function.

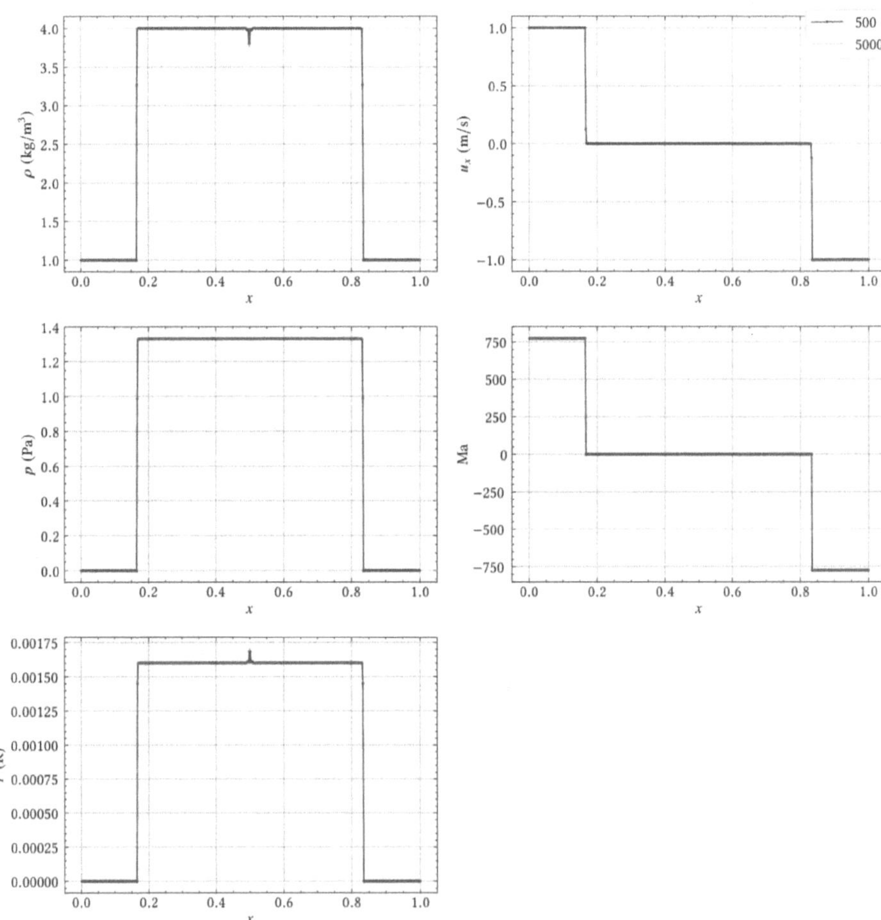

Figure 5.4: One-dimensional Noh problem at a final time of $t = 1$, using the Roe flux function.

5.3.2   Two-Dimensional Problems

In this section, a number of two-dimensional Euler problems are shown to demonstrate the DGH's order of accuracy for non-linear sets of PDEs and the efficacy of the AMR implementation. Additionally, some classical problems are presented for comparison against results shown previously in literature.

Ringleb's Flow

Ringleb's flow is a problem first described by Ringleb in 1940 [34], and can be used to determine the accuracy of a two-dimensional solver for the Euler equations, as an exact solution is available. This classical case describes a compressible gas flow between two stream lines. The flow is transonic and smooth, with a curved geometry used for the computational domain. The exact solution is obtained by defining the physical location of nodes used for the mesh in terms of the flow variables such as flow speed and angle. For this case, k is defined as a streamline parameter. After specifying values of $k_{\min} = 0.7$ for the outer wall of the geometry and $k_{\max} = 1.2$ for the inner wall, the velocity magnitude q is known to vary between $q_0 = 0.5$ and k. After iteratively solving for q, the speed of sound c, density ρ, and pressure p can be found with

$$c = \sqrt{1 - \frac{\gamma - 1}{2}q^2}, \quad \rho = c^{\frac{2}{\gamma-1}}, \quad p = \frac{1}{\gamma}c^{\frac{2\gamma}{\gamma-1}}. \tag{5.14}$$

The geometry is first defined with

$$x(q, k) = \frac{1}{2\rho}\left(\frac{2}{k^2} - \frac{1}{q^2}\right) - \frac{J}{2}, \quad y(q, k) = \pm\frac{1}{k\rho q}\sqrt{1 - \frac{q^2}{k^2}}, \tag{5.15}$$

where

$$J = \frac{1}{c} + \frac{1}{3c^3} + \frac{1}{5c^5} - \frac{1}{2}\ln\frac{1+c}{1-c}. \tag{5.16}$$

The Ringleb's mesh is then generated with uniform spacing in θ, which can be converted to q with $q = k\sin\theta$. In order to iteratively solve for q at a given point (x,y), fixed-point iteration is used. The isotachs (lines of constant speed) in the mesh are in the form of circles, defined by the equation

$$\left(x + \frac{J}{2}\right)^2 + y^2 = \left(\frac{1}{2\rho q^2}\right)^2. \tag{5.17}$$

Rearranging for q and applying fixed-point iteration yields

$$q^{k+1} = \sqrt{\frac{1}{2\rho^k\sqrt{\left(x + \frac{J^k}{2}\right)^2 + y^2}}}. \tag{5.18}$$

Here, q^{k+1} is the new value for q, and ρ^k, J^k are values calculated with the previous value q^k. An initial estimate of $q^0 = \frac{q_0 + k_{\min}}{2}$ is used, and new values of q^{k+1} are evaluated until a convergence

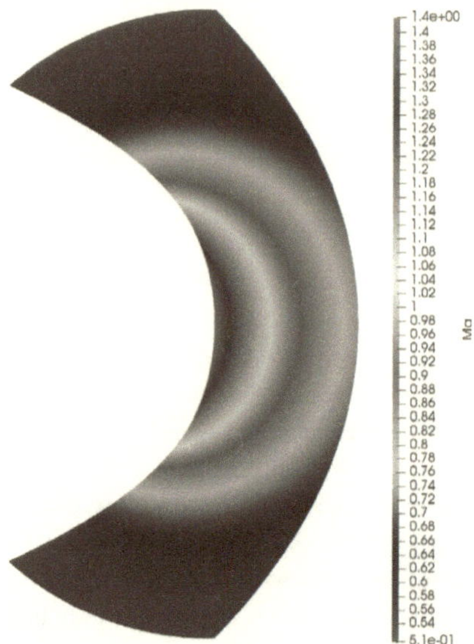

Figure 5.5: Exact solution to Ringleb's flow problem on a grid of 800 × 800 elements.

criterion of $|q^{k+1} - q^k| < \epsilon$ is met. In the present implementation, the convergence value is taken as $\epsilon = 1 \times 10^{-15}$. The exact solution is set as the initial condition over the entire domain, and is also imposed as boundary conditions on all edges of the mesh. The solution is then time marched until a sufficient residual reduction factor of 1×10^{-6} is met. The Roe flux function is used for this problem, with a CFL number of 0.3. Because the solution is smooth, no slope limiter is used for this case. The computed ℓ^2-norms of the error for an increasing number of elements N are shown in Table 5.2. As can be seen, third-order accuracy is sufficiently recovered on a non-Cartesian mesh for a set of non-linear PDEs.

Table 5.2: Convergence study for Ringleb's flow.

Number of Elements	ℓ^2 Error	Order
20 × 20	3.867×10^{-4}	—
40 × 40	8.470×10^{-5}	2.03
80 × 80	1.451×10^{-5}	2.45
100 × 100	7.962×10^{-6}	2.63
200 × 200	1.158×10^{-6}	2.74
400 × 400	1.593×10^{-7}	2.84
800 × 800	2.144×10^{-8}	2.88

Bump Case

For this case, supersonic inviscid flow over a bump is simulated. The domain is initially comprised of five blocks over the domain $x \in [-1\,\text{m}, 4\,\text{m}]$, $y \in [0\,\text{m}, 1\,\text{m}]$, with a circular arc located in the second block between $x = 0\,\text{m}$ and $x = 1\,\text{m}$. The far left boundary condition is held fixed as an inflow of air with properties $\rho = 1.225\,\text{kg/m}^3$, $p = 101.325\,\text{kPa}$ at a speed of Ma = 2, with a zero-gradient condition on the far right boundary and reflecting boundary conditions on the top and bottom. The solution is time-marched to $t = 0.02\,\text{s}$ using the HLLE flux function and a CFL number of 0.3. The initial mesh is comprised of five blocks, each of size 20×20 elements. Adaptive mesh refinement is then performed, and the solution is again time marched until steady state is met. This process of refining then time-marching to steady state is repeated four times, and the results are shown in Figure 5.6. The refinement criterion for this case calculated the ℓ^2-norm of the gradient of density within a block, and blocks with norm values greater than the arbitrary value of $20\,\text{kg/m}^2$ were refined. As can be seen, regions of high interest are refined several times over, while less notable areas are left un-refined.

Double Mach Reflection

This is a two-dimensional problem first described by Woodward and Colella [43], in which an ideal gas with constant $\gamma = 1.4$ is simulated from rest with initial conditions $W_0 = [1.4, 0, 0, 1]^\top$ on the domain $x \in [0, 3.25]$, $y \in [0, 1]$. A Mach 10 shock which propagates at an angle of 30° from the x-axis is initially set up at position $x = \frac{1}{6}$. A fixed time-dependent boundary condition is used on the top boundary, and a conditional reflecting condition is used on the bottom to allow the shock to continue to propagate during the simulation. The left boundary is held fixed with the Mach 10 shock state, and zero-gradient conditions are used on the right boundary. The simulation is time-marched to a final non-dimensional time of $t = 0.2$, at which point the shocks are near the right boundary of the domain. This problem is a good test for ensuring that all shocks propagate at the correct speeds, as the carbuncle phenomenon is often encountered when the Roe approximate Riemann solver is used for this case. The result shown in Figure 5.7 is obtained using the HLLE flux function and a CFL number of 0.3. Adaptive mesh refinement is performed every 500 iterations during time-marching, with the refinement criterion dependent on the ℓ^2-norm of the vorticity of a block. The domain is initially comprised of 1024 blocks, each of size 20×20 elements, with the final mesh containing 12 805 blocks for a total of 5 122 000 elements. The adaptive mesh refinement has done a good job here of tracking regions of interest, and coarsening the mesh back elsewhere. Waves seen behind the shock can be attributed to the fact that the scheme is not TVD, exacerbated slightly by less than frequent AMR.

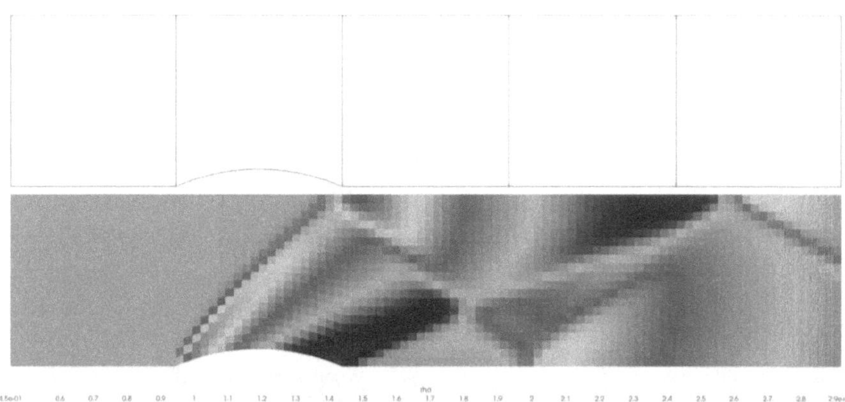

(a) Solution with zero levels of refinements.

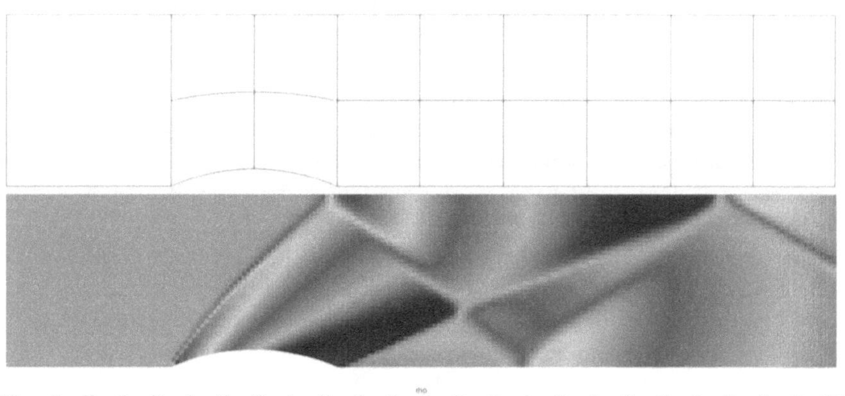

(b) Solution with one level of refinement.

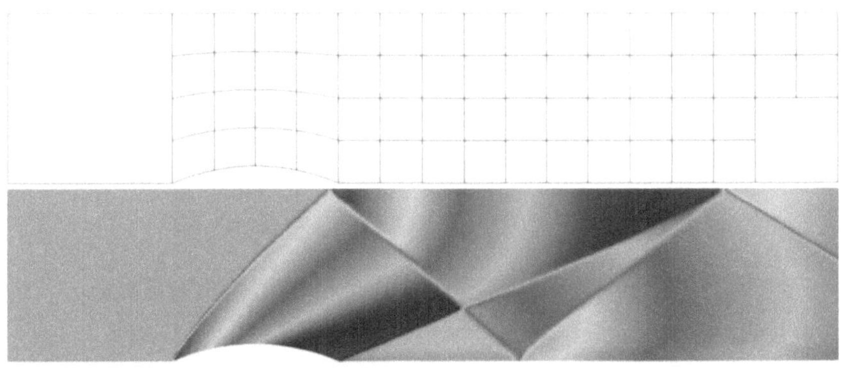

(c) Solution with two levels of refinements.

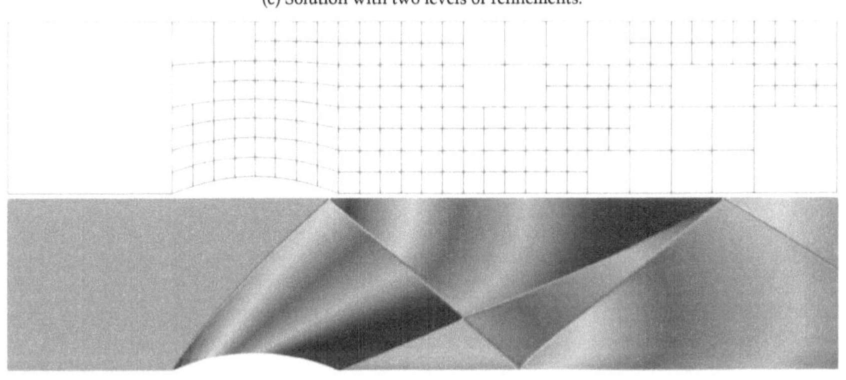

(d) Solution with three levels of refinements.

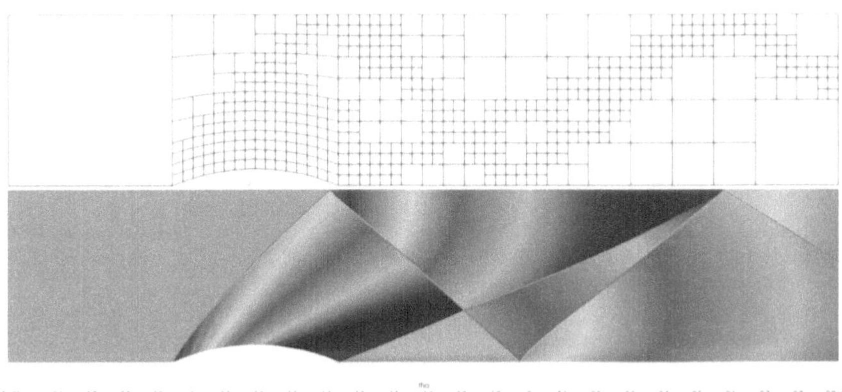

(e) Solution with four levels of refinements.

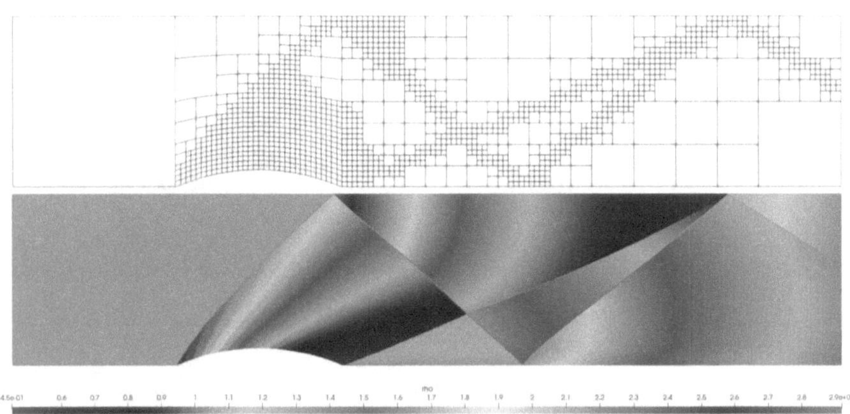

(f) Solution with five levels of refinements.

Figure 5.6: Solutions for Mach 2 flow over a bump at varying levels of refinement, obtained with a CFL number of 0.3 and the HLLE flux function.

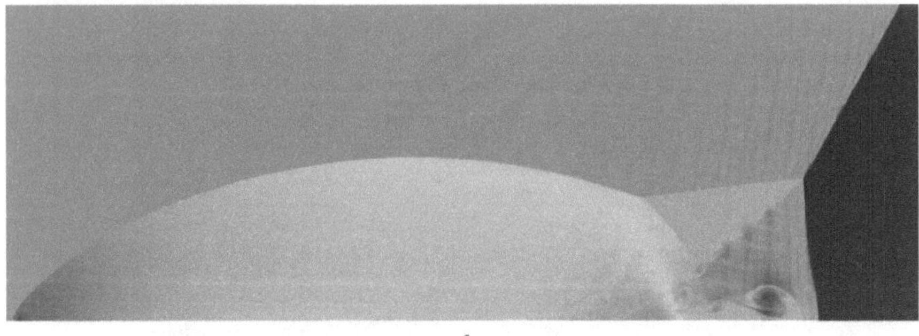

Figure 5.7: Solution for the double Mach reflection case at a final time of $t = 0.2$, using the HLLE flux function and a CFL number of 0.3.

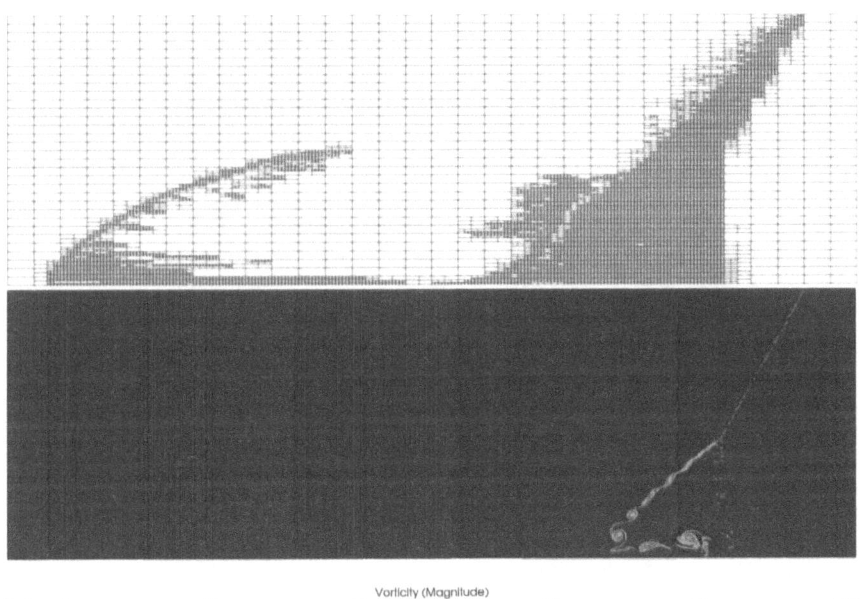

Figure 5.8: Outline plot for the structured mesh at the end of double Mach reflection simulation. A refinement criterion based upon the value of the ℓ^2-norm of vorticity in a block is used during time-marching.

Noh Problem

This is a classical two-dimensional problem first studied by Noh [30]. This problem once again uses an ideal gas with constant $\gamma = 5/3$ and initial conditions

$$W_0 = \begin{bmatrix} 1 \\ -\dfrac{x}{\sqrt{x^2+y^2}} \\ -\dfrac{y}{\sqrt{x^2+y^2}} \\ 1 \times 10^{-6} \end{bmatrix}. \tag{5.19}$$

It can be seen that the initial velocities (u_x, u_y) are directed towards the origin with a magnitude of 1. This problem has an exact solution for a given time t, as an infinitely strong symmetric circular shock is propagated from the origin. The exact solution is given by

$$W_{\text{Exact, behind shock}} = \begin{bmatrix} 16 \\ 0 \\ 0 \\ \frac{16}{3} \end{bmatrix}, \quad W_{\text{Exact, ahead of shock}} = \begin{bmatrix} 1 + \dfrac{t}{\sqrt{x^2+y^2}} \\ -\dfrac{x}{\sqrt{x^2+y^2}} \\ -\dfrac{y}{\sqrt{x^2+y^2}} \\ 1 \times 10^{-6} \end{bmatrix}. \tag{5.20}$$

The shock front is at radial position $r = t/3$, where t is the final time for the simulation. Reflecting boundary conditions are used on the lower x and y boundaries, while the exact solution is imposed on the upper bounds. The solution is then time-marched on a domain $(x, y) \in [0, 1]^2$ to a final time of $t = 2$. Results for this problem on a mesh of 400×400 elements are shown in Figure 5.9. The points above the exact solution on the right hand side of the shock in the scatter plot are a result of the carbuncle phenomenon, in line with the shock on the bottom and left boundaries. The numerical results obtained with the DGH scheme appear quite strong when compared to the results presented by Liska and Wendroff [27].

Rayleigh-Taylor Instability

The Rayleigh-Taylor instability is a physical phenomenon which occurs at the interface between two fluids with different densities, in which the heavier fluid is on top of the lighter fluid. For this problem, a gravitational source term is added to the two-dimensional Euler equations, such that

$$S = \begin{bmatrix} 0 \\ 0 \\ -\rho g_y \\ -\rho u_y g_y \end{bmatrix}. \tag{5.21}$$

Here, g_y is the gravitational acceleration. For this problem, a value of $g_y = 0.1$ is used. The problem is simulated on the bounds $(x, y) \in [-1/6, 1/6] \times [0, 1]$, with an interface between the

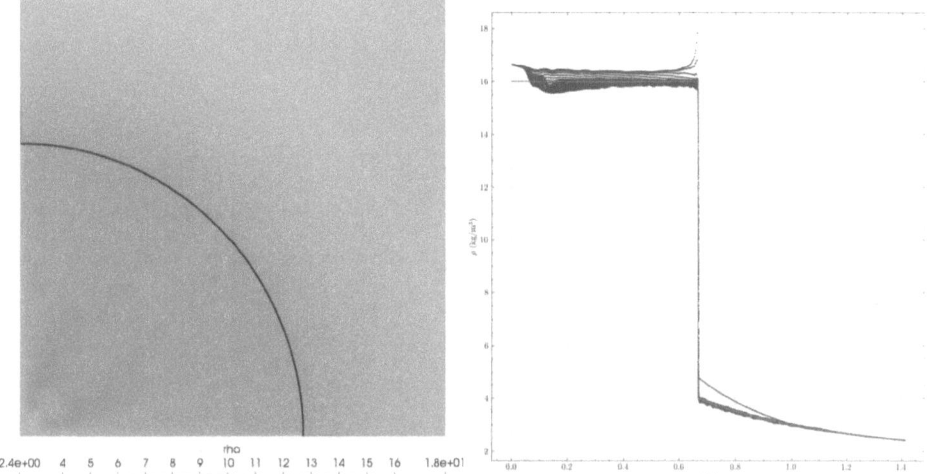

Figure 5.9: Results for the two-dimensional Noh problem at a final time of $t = 2$, using the HLLE flux function and a CFL number of 0.3. Density colormap is shown on the left, and a scatter plot of density versus radius is plotted on the right.

two fluids at $y_{\text{interface}} = \frac{1}{2} + 0.01\cos(6\pi x)$. The initial conditions for the bottom and top fluids are given in primitive form by

$$W_0 = \begin{cases} W_B, & y < y_{\text{interface}} \\ W_T, & y > y_{\text{interface}} \end{cases}, \quad W_B = \begin{bmatrix} 1 \\ 0 \\ 0 \\ p(y) \end{bmatrix}, \quad W_T = \begin{bmatrix} 2 \\ 0 \\ 0 \\ p(y) \end{bmatrix}, \quad (5.22)$$

where $p(y)$ is the hydrostatic pressure at a given value of y. Reflecting boundary conditions are used on all edges, and the solution is time-marched to a final time of $t = 8.5$. Figure 5.10 presents the solution obtained on a computational mesh of 100×400 elements, using the Roe flux function and a CFL number of 0.3. Because the solution is symmetric, it is shown mirrored about the y-axis. The DGH method captures the break-up of the interface between regions of high density and low density, demonstrating the low numerical dissipation of the scheme.

Kelvin-Helmholtz Instability

The Kelvin-Helmholtz instability occurs at the interface between two fluids with differing velocities and densities. For this problem, the instability is triggered at two interfaces, which are once again slightly perturbed lines at $(y_B, y_T) = (0.25, 0.75) \pm 0.01\cos(6\pi x)$. The initial conditions are

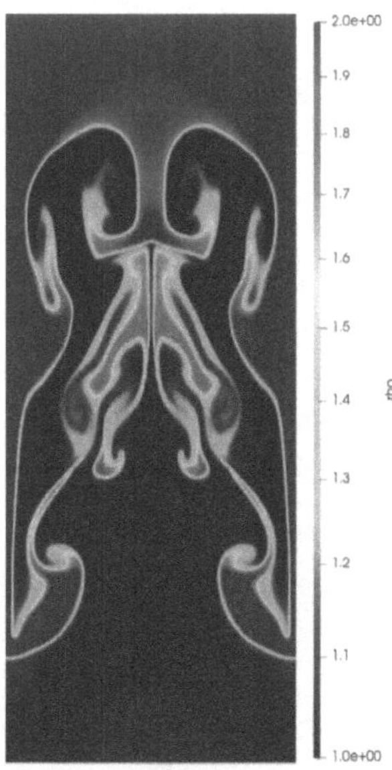

Figure 5.10: Solution for the Rayleigh-Taylor problem at time $t = 8.5$, using the Roe flux function and a CFL number of 0.3.

given by

$$W_0 = \begin{cases} W_1, & y_B < y < y_T \\ W_2, & \text{otherwise} \end{cases}, \quad W_1 = \begin{bmatrix} 2 \\ 0.5 \\ 0 \\ 2.5 \end{bmatrix}, \quad W_2 = \begin{bmatrix} 1 \\ -0.5 \\ 0 \\ 2.5 \end{bmatrix}. \tag{5.23}$$

The solution is time-marched to a final time of $t = 2$ on a domain $(x, y) \in [0, 1]^2$ with periodic boundary conditions in both directions. The solution shown in Figure 5.11 is obtained using 1024 blocks of size 200×200 cells, for a total of 40 960 000 elements. The scheme is efficiently able to maintain small-scale structures in the flow without showing excessive numerical dissipation.

5.3.3 Three-Dimensional Problems

In this section, the Euler equations are solved in three-dimensions to demonstrate the efficacy of the AMR algorithm.

Vorticity Generated by a Shock Wave

This is a three-dimensional case considered by Langseth and Leveque [21], in which shocks interact with regions of varying density. Initially, the gas is at rest everywhere with unity density and pressure, except for two cylindrical regions perpendicular to each other. The initial conditions

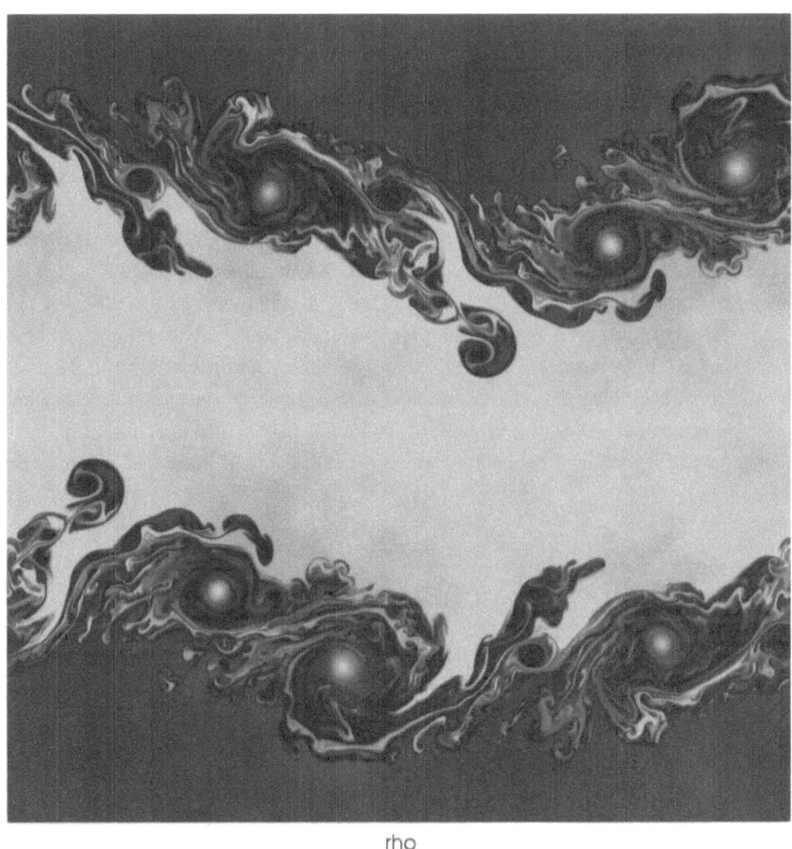

Figure 5.11: Solution for Kelvin-Helmholtz instability at time $t = 2$, using the Roe flux function and a CFL number of 0.3.

can be written mathematically as

$$W_0 = \begin{cases} W_A, & \sqrt{x^2 + y^2} < 0.2 \\ W_B, & \sqrt{(x - 0.4)^2 + z^2} < 0.2 \,, \\ W_C, & \text{otherwise} \end{cases} \quad W_A = \begin{bmatrix} 1 \\ 0 \\ 0 \\ 0 \\ 10 \end{bmatrix}, \; W_B = \begin{bmatrix} 0.1 \\ 0 \\ 0 \\ 0 \\ 1 \end{bmatrix}, \; W_C = \begin{bmatrix} 1 \\ 0 \\ 0 \\ 0 \\ 1 \end{bmatrix}. \tag{5.24}$$

The computational domain is $x \in [0, 1.5]$, $y \in [0, 1]$, $z \in [0, 0.5]$, and zero-gradient conditions are imposed on all boundaries. Initially, a parent block of size $20 \times 20 \times 20$ is refined into 512 child blocks. Adaptive mesh refinement is performed every 50 iterations, with a refinement criterion dependent on the ℓ^2-norm of the gradient of density. Time-marching is performed using the HLLE flux function and a CFL number of 0.3. When the front shock hits the low density cylinder, a large amount of vorticity is generated, and two "rolls" which rotate in opposite directions begin to propagate and interact. To display the structure of the solution, the value $|\vec{\nabla}\rho|$ is plotted over time. Contour plots for this value are presented in Figure 5.12, with outline plots for the computational mesh shown in Figure 5.13. As the rolls move across the domain, the adaptive mesh refinement accurately tracks the regions of high interest by coarsening the mesh behind the rolls and refining ahead of it. The quantity $|\vec{\nabla}\rho|$ is typically shown in Schlieren figures, and the numerical results obtained here are in agreement with the plots generated by Langseth and Leveque.

5.4 Shallow Water

The two-dimensional Shallow-Water equations can be written in balance law form as

$$\frac{\partial U}{\partial t} + \frac{\partial F_x}{\partial x} + \frac{\partial F_y}{\partial y} = S, \tag{5.25}$$

where

$$U = \begin{bmatrix} h \\ hu_x \\ hu_y \end{bmatrix}, \; F_x = \begin{bmatrix} hu_x \\ hu_x^2 + p \\ hu_x u_y \end{bmatrix}, \; F_y = \begin{bmatrix} hu_y \\ hu_x u_y \\ hu_y^2 + p \end{bmatrix}, \; S = \begin{bmatrix} 0 \\ -g\frac{\partial b}{\partial x} \\ -g\frac{\partial b}{\partial y} \end{bmatrix}. \tag{5.26}$$

Here, the pressure term can be written as $p = \frac{1}{2}gh^2$. Additionally, $b(x, y)$ is a function that defines the topography for the bottom of the domain.

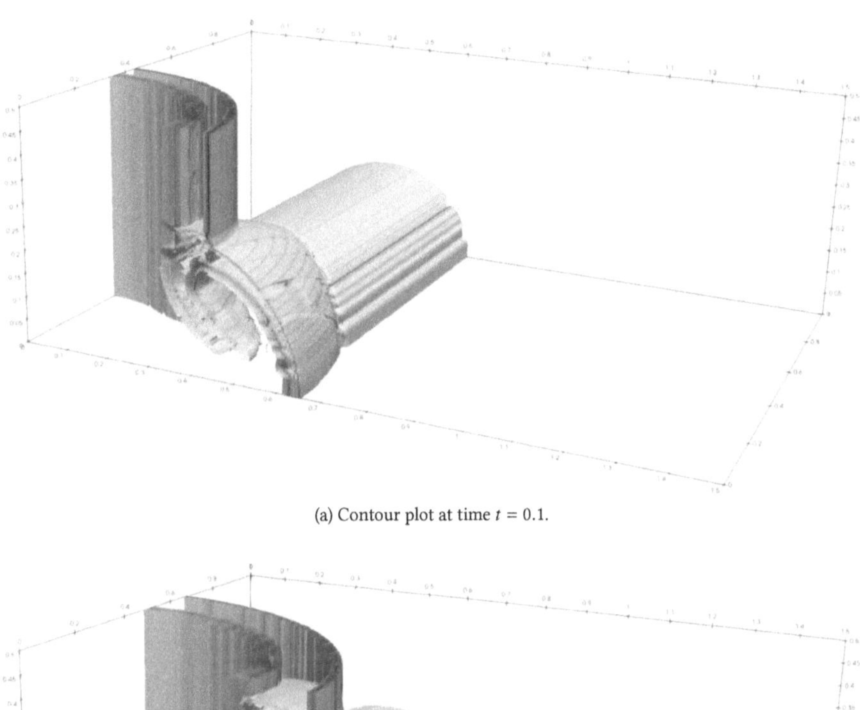

(a) Contour plot at time $t = 0.1$.

(b) Contour plot at time $t = 0.2$.

Figure 5.12: Contour plots of the value $|\vec{\nabla}\rho|$ for the three-dimensional vorticity generated by a shock wave case, using the HLLE flux function and a CFL number of 0.3.

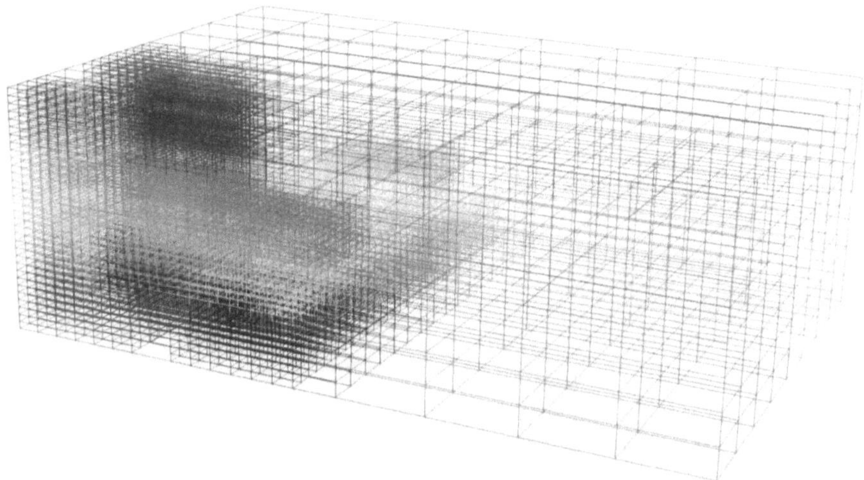

(a) Outline plot for vorticity case at time $t = 0.1$. At this point in time, the computational mesh is comprised of 6308 blocks.

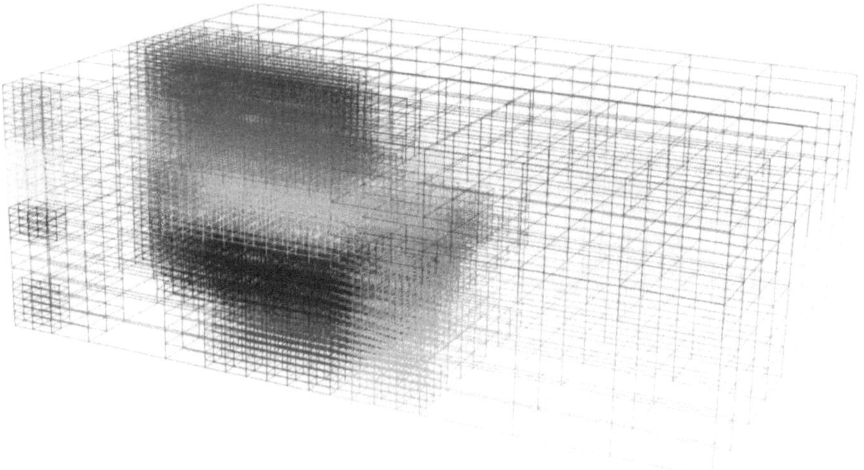

(b) Outline plot for vorticity case at time $t = 0.2$. At this point in time, the computational mesh is comprised of 8828 blocks.

Figure 5.13: Outline plots of the computational mesh for three-dimensional vorticity generated by a shock wave case.

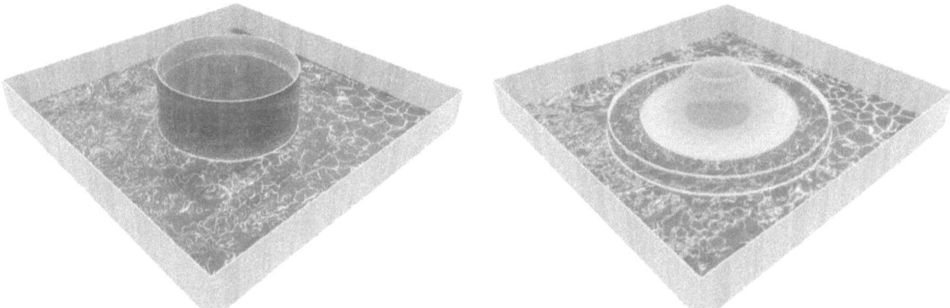

Figure 5.14: Initial condition and solution at time $t = 0.69$ for circular dam break problem, using the Roe flux function and a CFL number of 0.3.

Circular Dam Break

For this problem, water at rest $(u_x, u_y = 0)$ with an initial profile

$$h(x, y) = \begin{cases} 10, & r < 11 \\ 1, & \text{otherwise} \end{cases}, \tag{5.27}$$

where $r = \sqrt{(x - 25)^2 + (y - 25)^2}$ is simulated over a flat bottom, $b(x, y) = 0$, on the domain $(x, y) \in [0, 50]^2$. The solution is then time-marched to a final time of $t = 0.69$. The result obtained on a computational mesh of 400×400 elements using the Roe flux function and a CFL number of 0.3 is shown in Figure 5.14. As can be seen, the initial region of high water in the center of the domain has begun to disperse outwards, moving towards the reflecting boundaries.

5.5 Ten Moment Closure

The ten-moment (Gaussian) closure are obtained using the weight vector $W_{10} = \left[m, mv_i, mv_iv_j\right]^\top$. In two-dimensions, this yields a set of seven PDEs, and in three-dimensions, a set of ten. These PDEs are transport equations for the macroscopic properties ρ, u_i, and P_{ij}.

$$\frac{\partial \rho}{\partial t} + \frac{\partial \rho u_k}{\partial x_k} = 0, \tag{5.28a}$$

$$\frac{\partial}{\partial t}(\rho u_i) + \frac{\partial}{\partial x_k}(\rho u_i u_k + P_{ik}) = 0, \tag{5.28b}$$

$$\frac{\partial P_{ij}}{\partial t} + \frac{\partial}{\partial x_k}(u_k P_{ij}) + P_{jk}\frac{\partial u_i}{\partial x_k} + P_{ik}\frac{\partial u_j}{\partial x_k} = -\frac{1}{\tau}\left(P_{ij} - \frac{1}{3}P_{kk}\delta_{ij}\right). \tag{5.28c}$$

A significant weakness of the Gaussian closure is its inability to account for the effects of thermal diffusion, as the third-order velocity moments of the Gaussian are zero.

5.5.1 Two-Dimensional Problems

The two-dimensional ten moment (Gaussian) closure can be written in balance law form as

$$\frac{\partial U}{\partial t} + \frac{\partial F_x}{\partial x} + \frac{\partial F_y}{\partial y} = S, \tag{5.29}$$

where the vector of conserved variables, U, is expressed as

$$U = \begin{bmatrix} \rho \\ \rho u_x \\ \rho u_y \\ \rho u_x^2 + P_{xx} \\ \rho u_x u_y + P_{xy} \\ \rho u_y^2 + P_{yy} \\ P_{zz} \end{bmatrix}, \tag{5.30}$$

and the flux dyads F_x and F_y are given by

$$F_x = \begin{bmatrix} \rho u_x \\ \rho u_x^2 + P_{xx} \\ \rho u_x u_y + P_{xy} \\ \rho u_x^3 + 3u_x P_{xx} \\ \rho u_x^2 u_y + 2u_x P_{xy} + u_y P_{xx} \\ \rho u_x u_y^2 + u_x P_{yy} + 2u_y P_{xy} \\ u_x P_{zz} \end{bmatrix}, \quad F_y = \begin{bmatrix} \rho u_y \\ \rho u_x u_y + P_{xy} \\ \rho u_y^2 + P_{yy} \\ \rho u_x^2 u_y + 2u_x P_{xy} + u_y P_{xx} \\ \rho u_x u_y^2 + u_x P_{yy} + 2u_y P_{xy} \\ \rho u_y^3 + 3u_y P_{yy} \\ u_y P_{zz} \end{bmatrix}, \tag{5.31}$$

Finally, the source vector S has the form

$$S = -\frac{1}{\tau} \begin{bmatrix} 0 \\ 0 \\ 0 \\ \frac{2P_{xx}-P_{yy}-P_{zz}}{3} \\ P_{xy} \\ \frac{2P_{yy}-P_{xx}-P_{zz}}{3} \\ \frac{2P_{zz}-P_{xx}-P_{yy}}{3} \end{bmatrix}. \tag{5.32}$$

Here, the relaxation time τ is related to viscosity and hydrostatic pressure through

$$\tau = \frac{\mu}{p}, \tag{5.33}$$

where $p = P_{ii}/3$.

Stokes Flow

This is a classical simulation of low-Reynolds Stokes flow over a cylinder. The fluid simulated is Argon with a density of $\rho = 1.784\,\mathrm{kg/m^3}$, a free-stream velocity of $0.5\,\mathrm{m/s}$, and a pressure of $101\,325\,\mathrm{Pa}$. A constant viscosity of $\mu = 2.117 \times 10^{-5}\,\mathrm{Pa \cdot s}$ is used. The diameter of the cylinder is set to $1 \times 10^{-5}\,\mathrm{m}$, and the corresponding Reynolds and Knudsen numbers are $\mathrm{Re} = 0.421$, $\mathrm{Kn} = 0.0063$ respectively, with a Mach number of $\mathrm{Ma} = 0.0018$. An exact solution for this situation can be found using the Navier-Stokes equation, and is written using a two-dimensional stream function such that

$$\psi = \sin\theta \left(Ar^3 + Br \ln r + Cr + Dr^{-1} \right). \tag{5.34}$$

Here, A, B, C, and D are integration constants. This exact solution is imposed as the boundary condition on the far-field edge in the domain, and a no-slip condition is assumed at the cylinder wall. The free-stream velocity in the domain can be found in cylindrical coordinates as

$$v_\theta = -\frac{\partial \psi}{\partial r}, v_r = \frac{1}{r}\frac{\partial \psi}{\partial \theta}. \tag{5.35}$$

After determining the velocity field, the pressure and viscous-stress fields can be determined using the Navier-Stokes equations once more. For this simulation, a mesh comprised of 2080 blocks of size 50×50 elements each is used to represent the computational domain. The solution is then time-marched with a CFL number of 0.3 until a sufficient residual reduction criteria is met. The HLLE flux function is used as an approximate Riemann solver, and because the solution is smooth, no slope limiter is used for this case. The numerical results obtained are compared with the exact solution in Figure 5.15. Figure 5.15a shows the deviation of P_{xx} from the thermodynamic pressure p, which corresponds to the negative of the normal component of the viscous stress in the Navier-Stokes model. The numerical result is in strong agreement with the exact solution, though the stress magnitude is slightly under-predicted in front of and behind the cylinder. This is likely due to numerical error introduced by a large condition number as a result of extremely low-Mach number flow. The shear pressure P_{xy} is shown in Figure 5.15b. This value corresponds to the negative of the viscous shear stress τ_{xy} from the Navier-Stokes equations, and here the numerical result is nearly identical to the exact solution. Finally, the plot of the Mach number is shown in Figure 5.15c, which again shows strong agreement with the exact solution. The results here are impressive, considering the low speed of the flow, in addition to the fact no pre-conditioner is used for the wavespeeds of the system.

Crossing Beams with Acceleration

This is a two-dimensional problem originally studied by Vié et al. [41]. This case is an example of monodisperse flow, where two streams of particles cross.

The behaviour of multi-phase gas-particle flows can be characterized by the Stokes number, Stk. This non-dimensional value is defined as the ratio of the characteristic time of a particle to

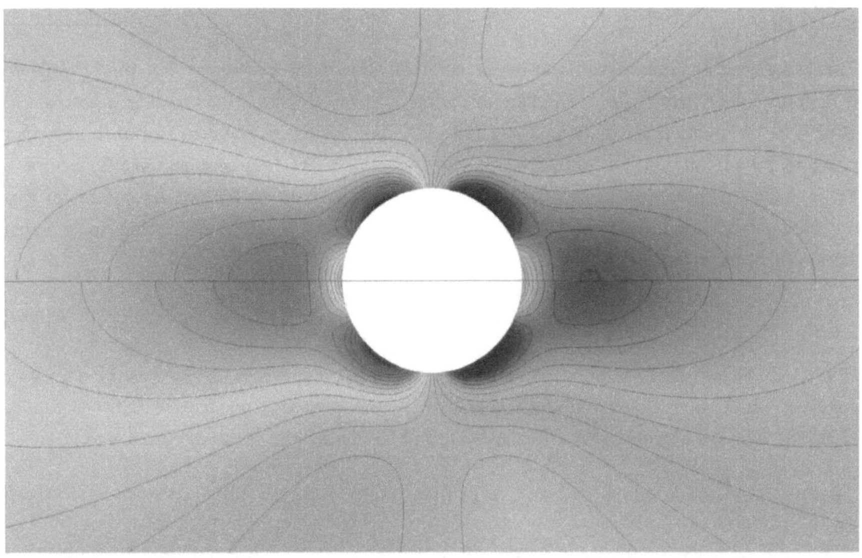

(a) Result for the deviatoric x-direction pressure $P_{xx} - p$ for Stokes Flow case.

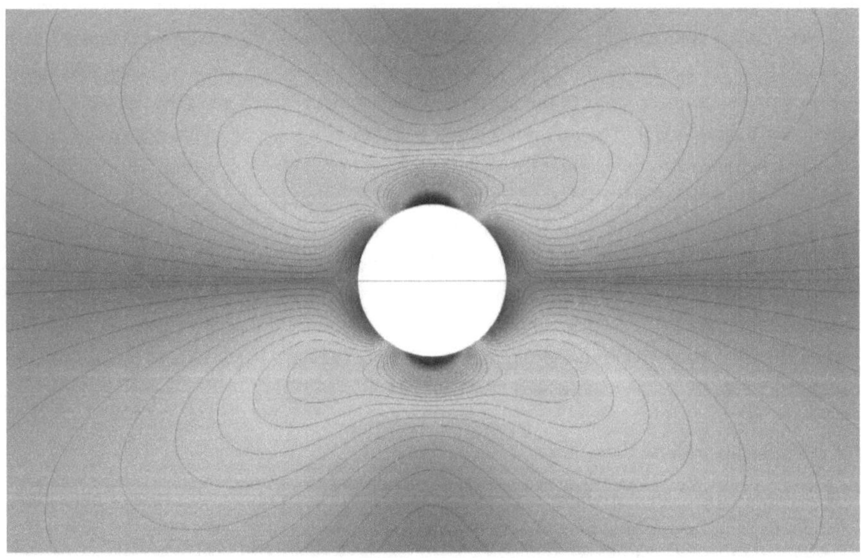

(b) Result for the shear pressure P_{xy} for Stokes Flow case.

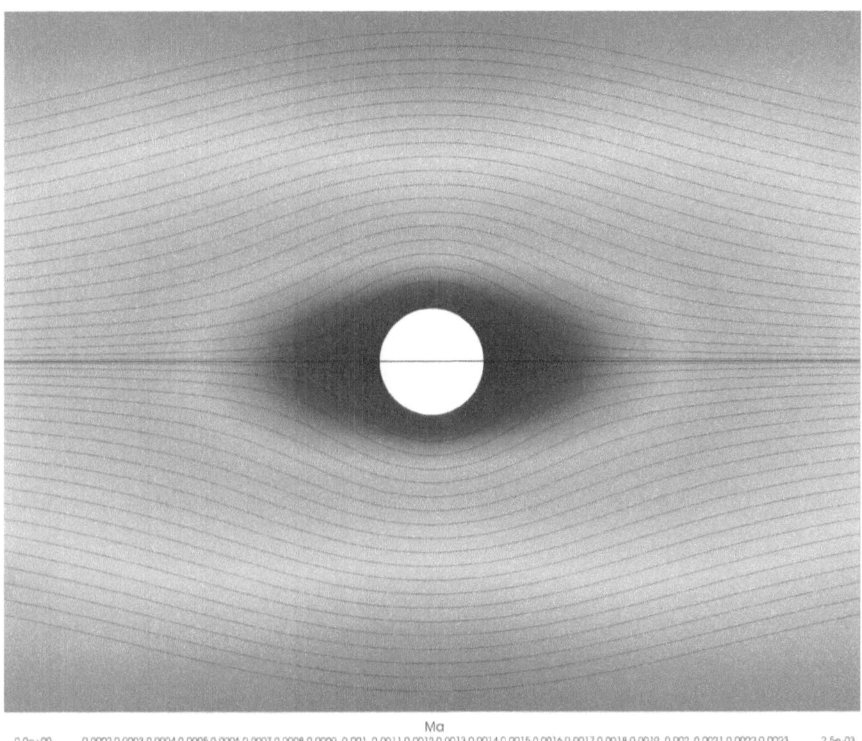

(c) Result for the Mach number for Stokes Flow case.

Figure 5.15: Results for the Stokes Flow case, using the HLLE flux function and a CFL number of 0.3. The numerical result is shown on the top half of the plot, while the exact solution is shown reflected in the lower half.

the characteristic time of the flow, such that

$$\text{Stk} = \frac{\tau |V_i|}{l},\tag{5.36}$$

where V_i is the velocity of the background flow, l is the characteristic length of the particle, and τ is the relaxation time due to drag forces. As the Stokes number becomes very large, traditional fluid mechanics cannot be used to accurately predict multi-phase flows. Multi-phase flows comprised of many particles moving at different velocities can be seen as similar to gas flows made up of atoms or molecules. As such, moment methods can provide a valid description of such flows.

For this case, the BGK operator is removed (thus ignoring particle collisions), and Stokes drag is instead used for the source term. The source vector for the two-dimensional Gaussian closure is then written as

$$S = \begin{bmatrix} 0 \\ \frac{\rho}{\tau}\left(u_{g_x} - u_x\right) \\ \frac{\rho}{\tau}\left(u_{g_y} - u_y\right) \\ \frac{2}{\tau}\left[\rho u_x \left(u_{g_x} - u_x\right) - P_{xx}\right] \\ \frac{1}{\tau}\left[\rho u_y \left(u_{g_x} - u_x\right) + \rho u_x \left(u_{g_y} - u_y\right) - 2P_{xy}\right] \\ \frac{2}{\tau}\left[\rho u_y \left(u_{g_y} - u_y\right) - P_{yy}\right] \\ -\frac{2}{\tau}P_{zz} \end{bmatrix}.\tag{5.37}$$

Here, $[u_{g_x}, u_{g_y}]$ are the background fluid velocities at the position of a particle, and τ is the relaxation time given by

$$\tau = \frac{\rho_p d^2}{18\mu_f},\tag{5.38}$$

where ρ_p is the density of the particle material, d is the particle diameter, and μ_f is the dynamic viscosity of the fluid. A value of $\tau = 5.0\,\text{s}$ is chosen for the current problem. This constant value was chosen such that particle trajectory crossing (PTC) will occur [41]. A velocity field is imposed for the background flow, such that

$$V_x = 0.2\,\text{m/s}, \; V_y = -\epsilon y.\tag{5.39}$$

Here, a value of $\epsilon = 1\,\text{s}^{-1}$ is used. The flow is constant in the x-direction, while providing a compressive acceleration field in the y-direction. The domain for this problem is chosen as $x \in [0\,\text{m}, 5\,\text{m}]$, $y \in [-1\,\text{m}, 1\,\text{m}]$, where the initial conditions are given as

$$W_0 = \begin{bmatrix} 1 \times 10^{-6}\,\text{kg/m}^3 \\ 0\,\text{m/s} \\ 0\,\text{m/s} \\ 1 \times 10^{-6}\,\text{Pa} \end{bmatrix}.\tag{5.40}$$

Two jets of particles enter the domain on the left hand boundary per the condition

$$W_L = \begin{cases} W_A, & -0.6 < y < -0.4 \\ W_0, & \text{otherwise} \end{cases}, \quad W_A = \begin{bmatrix} 1.0 \, \text{kg/m}^3 \\ 0.2 \, \text{m/s} \\ 0.0 \, \text{m/s} \\ 1.0 \times 10^{-6} \, \text{Pa} \end{bmatrix}, \quad (5.41)$$

while the other boundaries are held fixed. The exact solution for this case can be written by defining a Stokes number $\text{Stk} = \epsilon\tau$ and frequency $\omega^2 = \frac{1}{4}\left|\frac{1}{\tau^2} - \frac{4\epsilon}{\tau}\right|$. The critical Stokes number is given as $\text{Stk}_c = \frac{1}{4}$, and the solution is expressed as

$$y_p(t) = y_{p_0} \exp\left(-\frac{t}{2\tau}\right) \begin{cases} \exp(-\omega t), & \text{if } \text{Stk} \leq \text{Stk}_c \\ \cos(-\omega t) - \frac{1}{2\omega\tau}\sin(-\omega t), & \text{otherwise} \end{cases}. \quad (5.42)$$

The x-position of a particle is easily found with $x_p(t) = V_x t$, and the density of a beam at a given value of x is inversely proportional to the width of the beam. This problem is simulated on a computational mesh of 2048 blocks, each with 25×25 cells for a total of 1 280 000 elements. The solution is time marched to a final time of $t = 25$ s, using the HLL flux function and a CFL number of 0.3. The exact solution is shown in Figure 5.16a, and the numerical results obtained are shown in Figure 5.16b. This is a difficult problem to solve, as the density field is near zero everywhere but the location of the beams, and is singular at the regions where the beams cross. The solution obtained by the Gaussian closure does not quite allow for clean crossing of the beams, but the general structure of the exact solution is recovered. This is a known limitation of the ten-moment model, consistent with the results shown by Vié et al. [41].

5.6 Fourteen Moment Closure

The fourteen moment closure is the lowest order moment closure that has a treatment for heat transfer. It is generated with the weight vector $W_{14} = \left[m, mv_i, mv_i v_j, mv_i v^2, mv^4\right]^\top$ [28], and can be written in balance law form as

$$\frac{\partial U}{\partial t} + \frac{\partial F}{\partial x} = S, \quad (5.43)$$

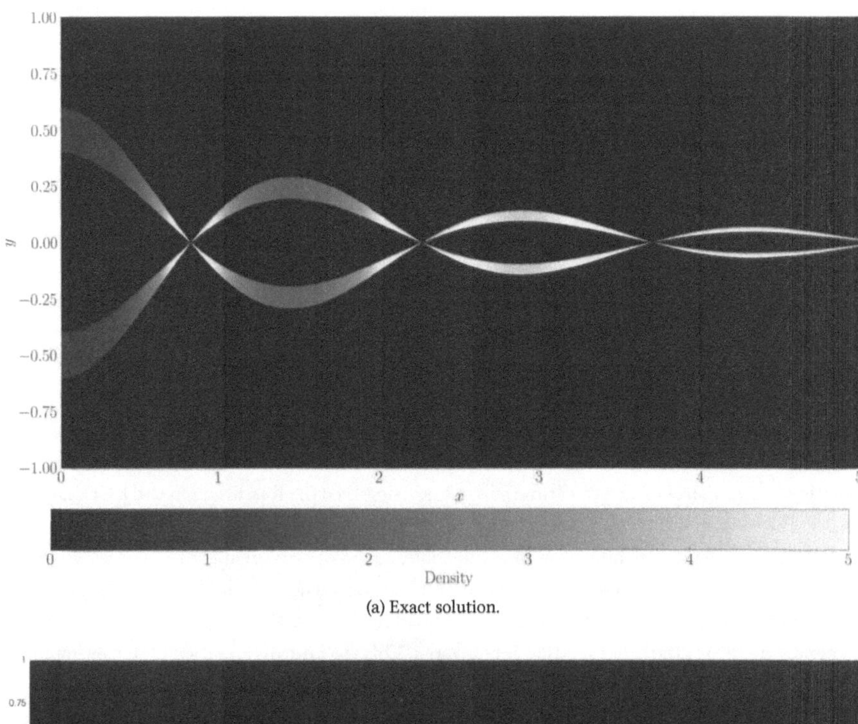

(a) Exact solution.

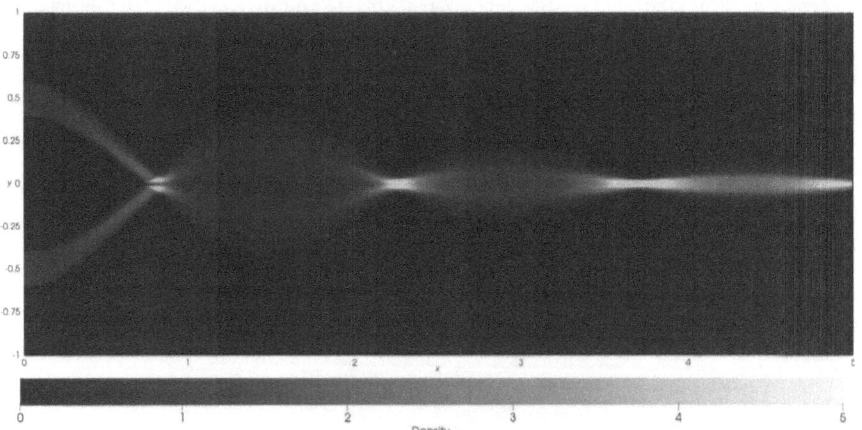

(b) Numerical solution, obtained using the HLL flux function and a CFL number of 0.3.

Figure 5.16: The exact solution and numerical solution for the crossing beams with acceleration case.

where

$$U = \begin{bmatrix} \rho \\ \rho u \\ \rho u^2 + P_{xx} \\ P_{rr} \\ \rho u^3 + u\left(3P_{xx} + P_{rr}\right) + Q_{xii} \\ \rho u^4 + u^2\left(6P_{xx} + 2P_{rr}\right) + 4uQ_{xii} + R_{iijj} \end{bmatrix},$$

$$F = \begin{bmatrix} \rho u \\ \rho u^2 + P_{xx} \\ \rho u^3 + 3uP_{rr} + Q_{xxx} \\ uP_{rr} + Q_{xrr} \\ \rho u^4 + u^2\left(6P_{xx} + 2P_{rr}\right) + 2u(Q_{xii} + Q_{xxx}) + R_{xxii} \\ \rho u^5 + u^3\left(10P_{xx} + 2P_{rr}\right) + u^2(6Q_{xii} + 4Q_{xxx}) + u\left(R_{iijj} + 4R_{xxii}\right) + S_{xiijj} \end{bmatrix}, \qquad (5.44)$$

$$S = \frac{1}{3\tau}\begin{bmatrix} 0 \\ 0 \\ P_{rr} - 2P_{xx} \\ 2P_{xx} - P_{rr} \\ 2u(P_{rr} - 2P_{xx}) - 3Q_{xii} \\ 4u^2\left(P_{rr} - 2P_{xx}\right) - 12uQ_{xii} + 5\frac{(P_{xx}+P_{rr})^2}{\rho} - 3R_{iijj} \end{bmatrix}.$$

Here, $P_{rr} = P_{yy}+P_{zz}$, $Q_{ijk} = \langle mc_ic_jc_k\mathcal{F}\rangle$ is the generalized heat-flux tensor, $R_{ijkl} = \langle mc_ic_jc_kc_l\mathcal{F}\rangle$ is a fourth-order moment, and $S_{ijklm} = \langle mc_ic_jc_kc_lc_m\mathcal{F}\rangle$ is a fifth-order moment. The closing relationships for the moment method outlined in McDonald's paper are given as [28]

$$Q_{xxx} = AQ_{xii}, \quad Q_{xrr} = (1 - A)Q_{xii}, \qquad (5.45)$$

$$R_{xxii} = \frac{A}{\sigma}\frac{Q_{xii}^2}{P_{xx}} + \frac{2(1-\sigma)P_{xx}^2 + (P_{xx} + P_{rr})P_{xx}}{\rho}, \qquad (5.46)$$

$$S_{xiijj} = \frac{A}{\sigma^2}\frac{Q_{xii}^3}{P_{xx}^2} + \frac{2}{\rho}\left(P_{xx} + P_{rr} + \left(1 - \sigma^{\frac{3}{5}}\right)\frac{P_{rr}^3 + 2P_{rr}^2P_{xx} + 24P_{xx}^3}{P_{rr}^2 + 6P_{xx}^2}\right)Q_{xii}, \qquad (5.47)$$

where

$$A = \frac{6P_{xx}^2}{P_{rr}^2 + 6P_{xx}^2}, \qquad (5.48)$$

and

$$\sigma = \frac{\left(2P_{xx}^2+P_{rr}^2\right)+(P_{xx}+P_{rr})^2-\rho R_{iijj}+\sqrt{\left[\left(2P_{xx}^2+P_{rr}^2\right)+(P_{xx}+P_{rr})^2-\rho R_{iijj}\right]^2+4\rho\left(2P_{xx}^2+P_{rr}^2\right)\frac{Q_{xii}^2}{P_{xx}}}}{2\left(2P_{xx}^2+P_{rr}^2\right)}. \tag{5.49}$$

Because there is no closed form for the fluxes of the system, this model provides numerical challenges and has restrictions imposed upon it to maintain global hyperbolicity.

Sod Shocktube

The first case is a classical Sod shocktube, using a collision time of $\tau = 1 \times 10^{-7}$ s. The initial conditions in primitive form are set to

$$\boldsymbol{W}_0 = \begin{cases} \boldsymbol{W}_L, & x < 5\,\mathrm{m} \\ \boldsymbol{W}_R, & x > 5\,\mathrm{m} \end{cases}, \quad \boldsymbol{W}_L = \begin{bmatrix} 4.696\,\mathrm{kg/m^3} \\ 0\,\mathrm{m/s} \\ 404\,400\,\mathrm{Pa} \end{bmatrix}, \quad \boldsymbol{W}_R = \begin{bmatrix} 1.408\,\mathrm{kg/m^3} \\ 0\,\mathrm{m/s} \\ 101\,100\,\mathrm{Pa} \end{bmatrix}. \tag{5.50}$$

The domain for this problem is $x \in [0\,\mathrm{m}, 10\,\mathrm{m}]$. The HLL flux function is used on a grid of 5000 elements, with a CFL number of 0.3. The solution shown in Figure 5.17 is in agreement with classical results, recovering the Euler solution due to use of a very stiff value for the relaxation time.

Impinging Jets

The second and third test cases are both impinging jets Riemann problems which use collision times of $\tau = 1000$ s. The initial conditions in primitive form are

$$\boldsymbol{W}_0 = \begin{cases} \boldsymbol{W}_L, & x < 5\,\mathrm{m} \\ \boldsymbol{W}_R, & x > 5\,\mathrm{m} \end{cases}, \quad \boldsymbol{W}_L = \begin{bmatrix} 1\,\mathrm{kg/m^3} \\ u_L \\ 100\,000\,\mathrm{Pa} \end{bmatrix}, \quad \boldsymbol{W}_R = \begin{bmatrix} 1\,\mathrm{kg/m^3} \\ u_R \\ 100\,000\,\mathrm{Pa} \end{bmatrix}. \tag{5.51}$$

For the second test case, $u_L = 200\,\mathrm{m/s}$, $u_R = -200\,\mathrm{m/s}$, and for the third test case, $u_L = 1000\,\mathrm{m/s}$, $u_R = -1000\,\mathrm{m/s}$. Both test cases were simulated at a CFL number of 0.3, using the HLL flux function, on the domain $x \in [0\,\mathrm{m}, 10\,\mathrm{m}]$ with zero-gradient boundary conditions. 5000 elements were used for each computation. The numerical results are shown in Figures 5.18 and 5.19. The density rises where the jets collide, with large values for the heat flux. Additionally, high pressure is seen all along the middle of the domain where the heat flux is zero.

5.7 Parallel Efficiency

As mentioned previously, the DGH method has been implemented within a framework parallelized with OMP and MPI. To prove the high parallel efficiency of the scheme, a strong scaling

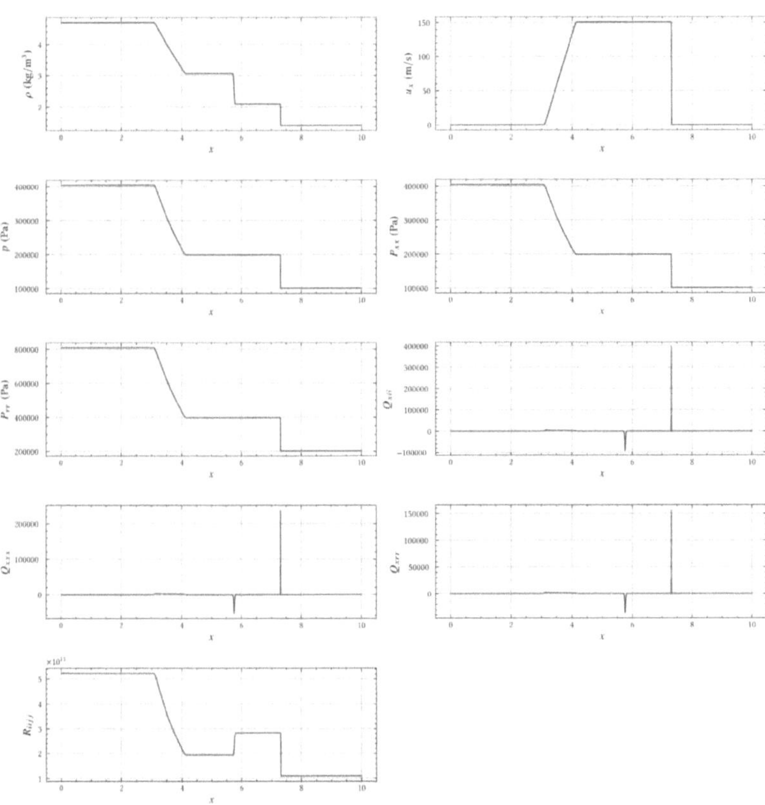

Figure 5.17: Solution for Sod shocktube case using the fourteen moment closure at a final time of $t = 5.0 \times 10^{-3}$ s.

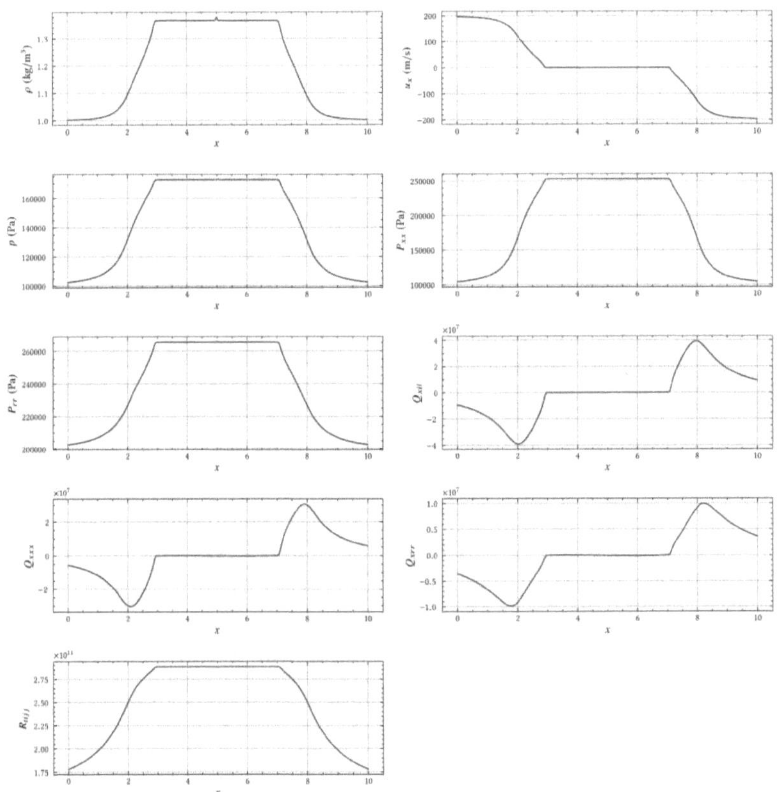

Figure 5.18: Solution for the first impinging jets case using the fourteen moment closure at a final time of $t = 5.0 \times 10^{-3}$ s.

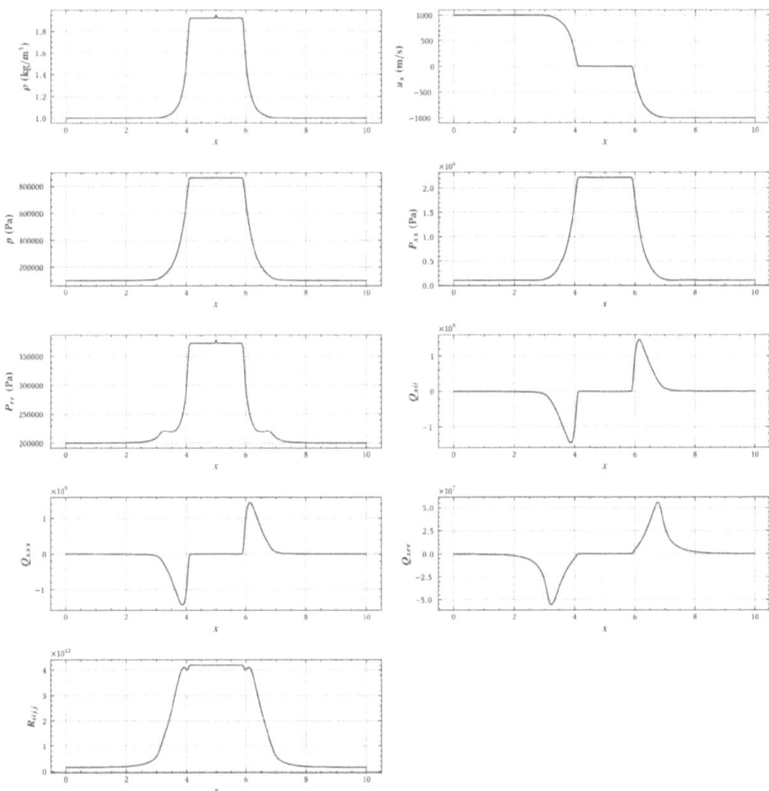

Figure 5.19: Solution for the second impinging jets case using the fourteen moment closure at a final time of $t = 5.0 \times 10^{-3}$ s.

analysis is performed. In a strong scaling analysis, the problem size is held constant as the number of computational cores is increased. The parallel speed-up S_n is defined as

$$S_n = \frac{t_1}{t_n}, \tag{5.52}$$

where t_1 is the wall-time of the computation on one node, and t_n is the wall-time of the computation on n nodes. The parallel efficiency E_n is then computed with

$$E_n = \frac{S_n}{n}. \tag{5.53}$$

For "perfect" scaling, the slope for S_n plotted against an increasing value of n is expected to be 1, and the efficiency E_n would be 100%.

All computations performed for this section were ran on the Niagara cluster provided by SciNet and Compute Canada, located at the University of Toronto. The cluster contains 2024 nodes, each with 40 Intel "Skylake" Xeon Gold 6148 processors with a clock speed of 2.4 GHz, or Intel "Cascadelake" Xeon Gold 6248 processors with a clock speed of 2.5 GHz. Each node has 202 GB of RAM. The parallel C++ framework was built with the Intel compiler, which boasted large speed-up over the GNU C++ compiler in some cases.

5.7.1 Three-Dimensional Euler Shockbox

The first strong scaling study is performed using a three-dimensional Euler shockcube case on the physical domain $(x, y, z) \in [-0.5\,\text{m}, 0.5\,\text{m}]^3$. The initial conditions in primitive form are given by

$$W_0 = \begin{cases} W_A, & x < 0\,\text{m}, y < 0\,\text{m}, z < 0\,\text{m} \\ W_B, & \text{otherwise} \end{cases},$$

$$W_A = \begin{bmatrix} 0.306\,25\,\text{kg/m}^3 \\ 0\,\text{m/s} \\ 0\,\text{m/s} \\ 0\,\text{m/s} \\ 25\,331.25\,\text{Pa} \end{bmatrix}, \quad W_B = \begin{bmatrix} 1.225\,\text{kg/m}^3 \\ 0\,\text{m/s} \\ 0\,\text{m/s} \\ 0\,\text{m/s} \\ 101\,325\,\text{Pa} \end{bmatrix}. \tag{5.54}$$

The solution is then time-marched to a final time of $7.5 \times 10^{-4}\,\text{s}$ with a CFL number of 0.3, the Venkatakrishnan slope limiter, and the HLLE flux function using 4096 blocks of size $20 \times 20 \times 20$ elements. Results for the study are shown in Table 5.3. Due to inter-CPU communication, perfect scaling is not quite achieved. Because a slope limiter is used for this case, an additional message has to be sent each time step containing the cell-centered values on boundary cells in addition to the always necessary message containing solution data on boundaries. These messages can serve as computational bottlenecks during time steps.

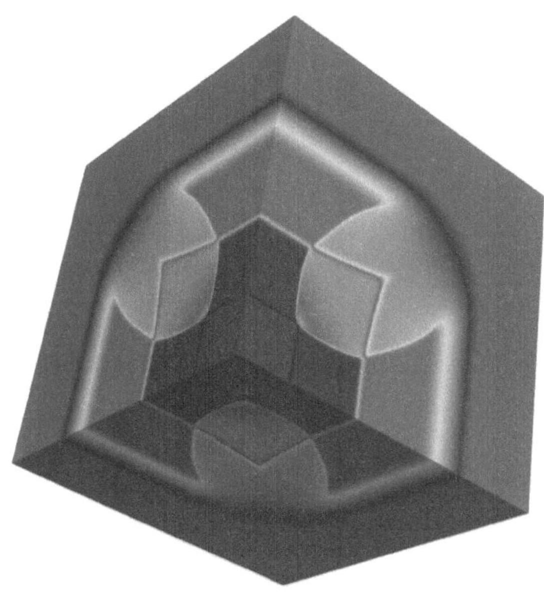

Figure 5.20: Solution for three-dimensional Euler shockcube case at time $t = 7.5 \times 10^{-4}$ s, using the Roe flux function and a CFL number of 0.3.

Table 5.3: Strong scaling study using three-dimensional Euler shockcube case with 4096 blocks of size $20 \times 20 \times 20$ elements.

# of CPUs	Wall-time (s)	Speed-up	Efficiency
32	13509.22	1.00	1.00
64	6436.74	2.10	1.05
128	3697.04	3.65	0.91
256	1815.43	7.44	0.93
512	934.10	14.46	0.90

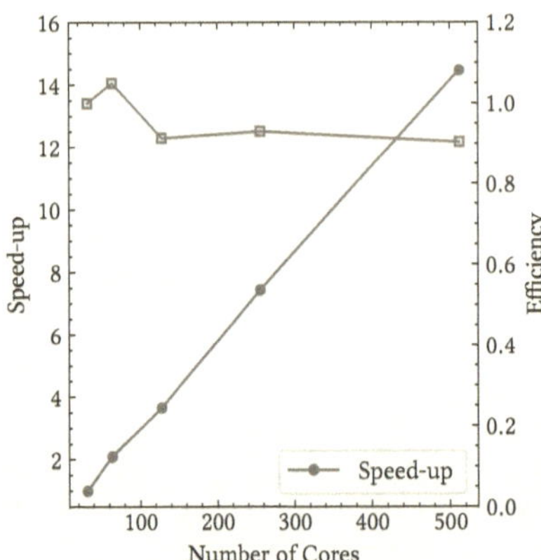

Figure 5.21: Plot for strong scaling study using three-dimensional Euler shockcube case with 4096 blocks of size $20 \times 20 \times 20$ elements.

Table 5.4: Strong scaling study using three-dimensional linear convection-relaxation case with 4096 blocks of size $20 \times 20 \times 20$ elements.

# of CPUs	Wall-time (s)	Speed-up	Efficiency
32	1094.53	1.00	1.00
64	547.16	2.00	1.00
128	278.09	3.94	0.98
256	140.51	7.79	0.97
512	71.61	15.28	0.96

5.7.2 Three-Dimensional Linear Convection-Relaxation

The case presented in Section 5.2 is used again for a strong scaling study. Due to the fact that no slope limiter is used for this case, it seems likely that the scaling will improve when compared to the results obtained with the three-dimensional Euler case. The mesh is initially refined into 4096 blocks of size $20 \times 20 \times 20$ elements and time-marched to the same final time using the same initial conditions, flux function, and CFL number as shown previously. The results for the study are shown in Table 5.4. As can be seen, the efficiency is closer to perfect in this study.

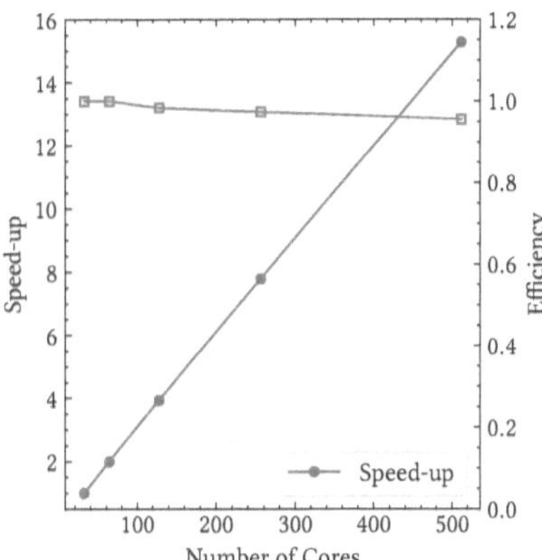

Figure 5.22: Plot for strong scaling study using three-dimensional linear convection-relaxation case with 4096 blocks of size $20 \times 20 \times 20$ elements.

Chapter 6

Conclusions

6.1 Summary

The current work presents the use of moment methods derived from the kinetic theory of gases for describing compressible gas flows both in and out of local thermodynamic equilibrium. The numerical advantages of these hyperbolic first-order PDEs are discussed, and a number of members of the maximum-entropy closure hierarchy are highlighted.

The formulation of the coupled space-time discontinuous-Galerkin Hancock scheme is shown in three dimensions. This scheme was designed specifically for the solution of hyperbolic-relaxation conservation laws with very stiff local source terms derived from moment closures. The scheme had not been implemented in a large-scale codebase up until now, and additionally had never been used for the solution of PDEs in more than one dimension. The implementation of block-based adaptive mesh refinement with isotropic refinement and coarsening is also discussed. The parallel framework is briefly described, tying together the implementation of the DGH scheme and adaptive mesh refinement algorithm.

Results obtained with the DGH scheme using AMR and load balancing on a modern distributed cluster for a number of different test cases and PDEs are presented. The accuracy of the scheme is proven to be third-order in three dimensions using the linear convection-relaxation equation. A large number of classical Euler cases are shown in one, two and three dimensions. Using the Ringleb's flow case, the scheme is proven to be third-order accurate on a non-Cartesian mesh for a set of non-linear PDEs. The shallow water equations are solved as an example of commonly used PDEs, further demonstrating the versatility of the scheme. The ten-moment Gaussian closure is shown to be accurate even without use of a pre-conditioner for low-speed viscous flows, and an example of multi-phase flow with Stokes drag is also presented. Results for some classical one-dimensional cases obtained using the fourteen-moment closure are shown. Finally, the parallel effiency of the implementation is studied using two strong scaling analyses, for up to hundreds of cores.

6.2   Future Work

The current implementations of the DGH scheme as well as the AMR algorithm are highly robust and extendable. Some advancements currently being worked towards include

- The implementation of derivative-driven, anisotropic refinement and coarsening for even greater dynamic accuracy and efficiency.

- Convergence studies for the DGH scheme on unstructured two-dimensional and three-dimensional meshes.

Some topics that are much larger in scope and would require extended research include

- The extension of the DGH scheme to fourth-order (or higher) temporal and spatial accuracy.

- The extension of the DGH scheme to second-order terms for the solution of PDEs with parabolic terms.